Leaves from

Gerard's Herball

arranged for Garden Lovers by

MARCUS WOODWARD

with 130 illustrations after the original woodcuts

DOVER PUBLICATIONS, INC.
NEW YORK

This Dover edition, first published in 1969, is an unabridged and unaltered republication of the work originally published by Gerald Howe, London, and Houghton Mifflin Company, Boston, in 1931.

The work upon which the present volume is based, *The Herball or General Historie of Plantes* by John Gerard, was originally published in·1597. A second edition enlarged and amended by Thomas Johnson was published in 1633, and reprinted in 1636.

Standard Book Number: 486-22343-4
Library of Congress Catalog Card Number: 79-82793

Manufactured in the United States of America
Dover Publications, Inc.
180 Varick Street
New York, N.Y. 10014

T HE very friendly reception accorded to *Gerard's Herball: The Essence thereof*, published in 1927, and now out of print, has suggested that there may be a place for a new edition, popular in form and price, yet no mere anthology but a full book such as all lovers of herbs and gardens would wish to have at hand at every season of the year. By compressing the text and omitting Johnson's comments and all cross headings, it has been found possible not only to preserve all the most characteristic passages, full of Gerard's 'sly humour and well-flavoured English', given before, but to add several chapters of great interest and delight, notably those which discourse of trees. Then the entire book has been rearranged so as to form as it were a garden calendar, the plants being grouped according to the time of their flowering or especial appeal. (Some allowance must be made for climate, and still more for the vagaries of our Author). For the most part the flowers and shrubs are the favourites of Gerard's day, and of ours too, in spite of the botanical discoveries of three hundred years. But who would be so heartless as to exclude the Barnacle Goose, and some few others that are unlikely to be grown with success in our cold climate and with cold reason?

For the full story of the Herball and its sources, together with the Life of Gerard, the reader is referred to the 1927 volume. The editor has however included some notes to help in identifying the varieties described, and there has been added a brief table of some of the more important ' vertues '.

TABLE OF CONTENTS

TO THE RIGHT HONORABLE HIS SINGULAR GOOD LORD & MASTER, SIR WILLIAM CECIL

Knight, Baron of Burghley, Master of the Court of
Wards & Liveries, Chancellor of the Universitie
of Cambridge, Knight of the most noble Order
of the Garter, one of the Lords of her Majesties
most honorable Privy Councell, and Lord
high Treasurer of *England.*

AMONG the manifold creatures of God (right
Honorable, and my singular good Lord) that have
all in all ages diversly entertained many excellent wits,
and drawn them to the contemplation of the divine
wisdome, none have provoked mens studies more, or
satisfied their desires so much as plants have done, and
that upon just and worthy causes: for if delight may
provoke mens labor, what greater delight is there than
to behold the earth apparelled with plants, as with a
robe of embroidered worke, set with Orient pearles and
garnished with great diversitie of rare and costly jewels?
If this varietie and perfection of colours may affect the
eie, it is such in herbs and floures, that no *Apelles*, no
Zeuxis ever could by any art expresse the like: if odours
or if taste may worke satisfaction, they are both so
soveraigne in plants, and so comfortable that no confec-
tion of the Apothecaries can equall their excellent vertue.
But these delights are in the outward senses: the principal
delight is in the mind, singularly enriched with the

knowledge of these visible things, setting forth to us the invisible wisdome and admirable workmanship of Almighty God. The delight is great, but the use greater, and joyned often with necessitie. In the first ages of the world they were the ordinary meate of men, and have continued ever since of necessary use both for meates to maintaine life, and for medicine to recover health. The hidden vertue of them is such, that (as *Pliny* noteth) the very bruit beasts have found it out: and (which is another use that he observes) from thence the Dyars tooke the beginning of their Art.

Furthermore, the necessary use of those fruits of the earth doth plainly appeare by the great charge and care of almost all men in planting & maintaining of gardens, not as ornaments onely, but as a necessarie provision also to their houses. And here beside the fruit, to speake againe in a word of delight, gardens, especialy such as your Honor hath, furnished with many rare Simples, do singularly delight, when in them a man doth behold a flourishing shew of Summer beauties in the midst of Winters force, and a goodly spring of flours, when abroad a leafe is not to be seene. Besides these and other causes, there are many examples of those that have honoured this science: for to passe by a multitude of the Philosophers, it may please your Honor to call to remembrance that which you know of some noble Princes, that have joyned this study with their most important matters of state: *Mithridates* the great was famous for his knowledge herein, as *Plutarch* noteth. *Euax* also King of Arabia, the happy garden of the world for principall Simples, wrot of this argument, as *Pliny* sheweth. *Dioclesian* likewise, might have had his praise, had he not drowned all his honour in the bloud of his persecution. To conclude this point, the example of *Solomon* is before the rest, and greater, whose wisdome and knowledge was such, that hee was able to set out the nature of all plants from the highest Cedar to the lowest Mosse. But my very good Lord, that which sometime was the study of great Phylosophers and mightie Princes, is now neglected, except it be of some few, whose spirit and wisdome hath carried them among other parts of wisdome and counsell, to a care and studie of speciall herbes both for the furnishing of their gardens, and furtherance of their knowledge: among whom I may justly affirme and publish your Honor to be one, being my selfe one of your servants, and a long time witnesse thereof: for under your Lordship I have served, and that way emploied my principall study and almost all my time, now by the space of twenty yeares. To the large and singular furniture of this noble Island I have added from forreine places all the varietie of herbes and floures that I might any way obtaine, I have laboured with the soile to make it fit for plants, and with the plants, that they might delight in the soile, that so they might live and prosper under our clymat, as in their native and proper countrey: whát my successe hath beene, and what my furniture is, I leave to the report of they that have seene your Lordships gardens, and the little plot of myne owne especiall care and husbandry. But because gardens are privat, and many times finding an ignorant or a negligent successor, come soone to ruine, there be that have sollicited me, first by my pen, and after by the Presse to make my labors common, and to free them from the danger wherunto a garden is subject: wherein when I was overcome, and had brought this History or report of the nature of Plants to a just volume, and had made it (as the Reader may by comparison see) richer than former Herbals, I found it no question unto whom I might dedicate my labors; for considering your good Lordship, I found none of whose favour and goodnes I might sooner presume, seeing I have found you ever my very good Lord and Master. Againe, considering my duty and your Honors merits, to whom may I better recommend my Labors, than to him unto whom I owe my selfe, and all that I am able in your service or devotion to performe? Therefore under hope of your Honorable and accustomed favor I present this Herball to your Lordships protection; and not as an exquisite Worke (for I know my meannesse) but as the greatest gift and chiefest argument of duty that my labour and service can affoord: wherof if there be no other fruit, yet this is of some use, that I have ministred Matter for Men of riper wits and deeper judgements to polish, and to adde to my large additions where any thing is defective, that in time the Worke may be perfect. Thus I humbly take my leave, beseeching God to grant you yet many daies to live to his glory, to the support of this State under her Majestie our dread Soveraigne, and that with great increase of honor in this world, and all fulnesse of glory in the world to come.

Your Lordships most humble

and obedient Servant,

JOHN GERARD.

THE HERBAL

BULBOUS VIOLETS

The bulbous Violet riseth out of the ground, with two small leaves flat and crested, of an overworne greene colour, betweene the which riseth up a small and tender stalke of two hands high; at the top whereof commeth forth of a skinny hood a small white floure of the bignesse of a Violet, compact of six leaves, three bigger, and three lesser, tipped at the points with a light greene; the smaller are fashioned into the vulgar forme of an heart, and prettily edged about with green; the other three leaves are longer, and sharpe pointed. The whole floure hangeth downe his head, by reason of the weake foot-stalke whereon it groweth. The root is small, white, and bulbous.

Bulbous Violet

Some call them also Snow drops. This name *Leucoium*, without his Epithite *Bulbosum*, is taken for the Wall-floure, and stocke Gillofloure, by all moderne Writers.

Touching the faculties of these bulbous Violets we have nothing to say, seeing that nothing is set downe hereof by the antient Writers, nor any thing observed by the moderne; onely they are maintained and cherished in gardens for the beautie and rarenesse of the floures, and sweetnesse of their smell.

Violets

The Violets called the blacke or purple violets, or March Violets of the garden, have a great prerogative about others, not only because the mind conceiveth a certain pleasure and recreation by smelling and handling those most odoriferous floures, but also for that very many by these violets receive ornament and comely grace; for there be made of them garlands for the head, nosegaies and poesies, which are delightfull to looke on and pleasant to smel to, speaking nothing of their appropriat vertues; yea gardens themselves receive by these the greatest ornament of all, chiefest beauty, and most excellent grace, and the recreation of the minde which is taken hereby cannot be but very good and honest; for they admonish and stirre up a man to that which is comely and honest; for floures through their beauty, variety of colour, and exquisit forme, do bring to a liberall and gentle manly minde, the remembrance of honestie, comlinesse, and all kindes of vertues: for it would be an unseemly and filthy thing (as a certain wise man saith) for him that doth looke upon and handle faire and beautiful things, to have his mind not faire, but filthy and deformed.

The blacke or purple Violet doth forthwith bring from the root many leaves, broad, sleightly indented in the edges, rounder than the leaves of Ivy; among the midst wherof spring up fine slender stems, and upon every one a beautifull flour sweetly smelling, of a blew darkish purple, consisting of five little leaves, the lowest whereof is the greatest: after them do appeare little hanging cups or knaps, which when they be ripe do open and divide themselves into three parts. The seed is smal, long, and somwhat round withall: the root consisteth of many threddy strings.

The white garden Violet hath many milke white floures, in forme and figure like the precedent; the colour of whose floures especially setteth forth the difference.

The double garden Violet hath leaves, creeping branches, and roots like the garden single Violet; differing in that, that this Violet bringeth forth most beautifull sweet double floures, and the other single.

The white double Violet likewise agrees with the other of his kinde, differing onely in the colour; for as the last described bringeth double blew or purple flours, contrariwise this plant beareth double white floures, which maketh the difference.

The Violet is called in Greeke, *Ion*: in Latine, *Nigra viola* or blacke Violet, of the blackish purple colour of the floures. The Apothecaries keepe the Latine name *Viola*, but they call it *Herba Violaria*, and *Mater Violarum*: in Spanish, *Violeta*: in English, Violet. *Nicander* beleeveth that the Grecians did call it *Ion*, because certain Nymphs of Iönia gave that floure first to *Jupiter*. Others say it was called because when *Jupiter* had turned the yong damosell *Iö*, whom he tenderly loved, into a Cow, the earth brought forth this floure for her food; which being made for her sake, received the name from her: and thereupon it is thought that the Latines also called it *Viola*, as though they should say *Vitula*, by blotting out the letter *t*.

Violet

The floures are good for all inflammations, especially of the sides and lungs; they take away the hoarsenesse of the chest, the ruggednesse of the winde-pipe and jawes, and take away thirst.

There is likewise made of Violets and sugar certaine plates called Sugar violet, Violet tables, or Plate, which is most pleasant and wholesome, especially it comforteth the heart and the other inward parts.

SPRING SAFFRON

Wilde Saffron hath small short grassie leaves, furrowed or channelled downe the midst with a white line or streak: among the leaves rise up small floures in shape like unto the common Saffron, but differing in color; for this hath floures of mixt colors; that is to say, the ground of the floure is white, stripped upon the backe with purple, and dasht over on the inside with a bright shining murrey color; the other not. In the middle of the floures come forth many yellowish chives, without any smell of Saffron at all. The root is small, round, and covered with a browne skin of filme like unto the roots of common Saffron.

We have likewise in our London gardens another sort, like unto the other wilde Saffrons in every point, saving that this hath floures of a most perfect shining yellow colour, seeming a far off to be a hot glowing cole of fire.

There is found among Herbarists another sort, not differing from the others, saving that this hath white floures, contrary to all the rest.

Lovers of plants have gotten into their gardens one sort hereof with purple or Violet coloured flours, in other respects like unto the former.

All these wild Saffrons we have growing in our London gardens.

DAFFODILS

The first of the Daffodils is that with the purple crowne or circle, having small narrow leaves, thicke, fat, and full of slimie juice; among the which riseth up a naked stalke smooth and hollow, of a foot high, bearing at the top a faire milke white floure growing forth of a hood or thin filme such as the flours of onions are wrapped in: in the midst of which floure is a round circle or small coronet of a yellowish colour, purfled or bordered about the edge of the said ring or circle with a pleasant purple colour; which being past, there followeth a thicke knob or button, wherein is contained blacke round seed. The root is white, bulbous or Onion-fashion.

The second kind of Daffodill is that sort of *Narcissus* or Primrose peerelesse that is most common in our country gardens, generally knowne everie where. It hath long fat and thick leaves, full of a slimie juice; among which riseth up a bare thicke stalke, hollow within and full of juice. The floure groweth at the top, of a yellowish white colour, with a yellow crowne or circle in the middle, and floureth in the moneth of Aprill, and sometimes sooner. The root is bulbous fashion.

There are three or foure reflex *Junquilia's*, whose cups hang downe, and the six incompassing leaves turne up or backe, whence they take their names.

The Daffodils with purple coronets grow wilde in sundry places, chiefly in Burgondie, and in Suitzerland in medowes.

Theocritus affirmeth the Daffodils to grow in medowes, in his 19 Eidyl, or 20 according to some editions: where he writeth, That the faire Lady *Europa* entring with her Nymphs into the medowes, did gather the sweet smelling daffodils; in these verses:

The reflex Junquilia

5

Which we may English thus:

> But when the Girles were come into
> The medowes flouring all in sight,
> That Wench with these, this Wench with those
> Trim floures, themselves did all delight:
> She with the Narcisse good in sent,
> And she with Hyacinths content.

But it is not greatly to our purpose, particularly to seeke out their places of growing wilde, seeing we have them all & everie one of them in our London gardens, in great aboundance. The common wilde Daffodill groweth wilde in fields and sides of woods in the West parts of England.

Galen saith, That the roots of Narcissus have such wonderfull qualities in drying, that they consound and glew together very great wounds, yea and such gashes or cuts as happen about the veins, sinues, and tendons. They have also a certaine clensing facultie.

The root of Narcissus stamped with hony and applied plaister-wise, helpeth them that are burned with fire, and joineth together sinues that are cut in sunder. Being used in manner aforesaid it helpeth the great wrenches of the ancles, the aches and pains of the joints. The same applied with hony and nettle seed helpeth Sun burning. Being stamped with the meale of Darnel and hony, it draweth forth thorns and stubs out of any part of the body.

SOW-BREAD

The common kinde of Sow-bread, called in shops *Panis porcinus*, and *Arthanita*, hath many greene and round leaves like unto Asarabacca, saving that the upper part of the leaves are mixed here and there confusedly with white spots, and under the leaves next the ground of a purple colour: among which rise up little stemmes like unto the stalkes of violets, bearing at the top small purple floures, which turne themselves backward (being full blowne) like a Turks cap, or Tulepan, of a small sent or savour, or none at all: which being past, there succeed little round knops or heads which containe slender browne seeds: these knops are wrapped after a few daies in the small stalkes, as thred about a bottome, where it

Sow-bread

remaineth so defended from the injurie of Winter close upon the ground, covered also with the greene leaves aforesaid, by which meanes it is kept from the frost, even from the time of his seeding, which is in September, untill June: at which time the leaves doe fade away, the stalkes & seed remaining bare and naked, whereby it injoyeth the Sun (whereof it was long deprived) the sooner to bring them unto maturitie.

Sow-bread groweth plentifully about Artoies and Vermandois in France, and in the Forest of Arden, and in

Brabant. It is reported unto mee by men of good credit, that *Cyclamen* or Sow-bread groweth upon the mountaines of Wales; on the hils of Lincolnshire, and in Somerset shire. Being beaten and made up into trochisches, or little flat cakes, it is reported to be a good amorous medicine to make one in love, if it be inwardly taken.

Muscari, or Musked Grape-floure

Yellow Muscarie hath five or six long leaves spread upon the ground, thicke, fat, and full of slimie juyce, turning and winding themselves crookedly this way and that way, hollowed alongst the middle like a trough, as are those of faire haired Jacinth, which at the first budding or springing up are of a purplish colour; but being growne to perfection, become of a darke greene colour; amongst the which leaves rise up naked, thicke, and fat stalkes, infirme and weake in respect of the thicknesse and greatnesse thereof, lying also upon the ground as do the leaves; set from the middle to the top on every side with many yellow floures, every one made like a small pitcher or little box, with a narrow mouth, exceeding sweet of smell like the savour of muske, whereof it tooke the name *Muscari*. The seed is closed in puffed or blowne up cods, confusedly made without order, of a fat and spongeous substance, wherein is contained round blacke seed. The root is bulbous or onion fashion, whereunto are annexed certaine fat and thicke strings like those of Dogs-grasse.

These plants came from beyond the Thracian Bosphorus, out of Asia, and from about Constantinople, and by the means of Friends have been brought into these parts of Europe, whereof our London gardens are possessed.

There hath not as yet any thing beene touched concerning the nature or vertues of these Plants, onely they are kept and maintained in gardens for the pleasant smell of their floures, but not for their beauty, for that many stinking field floures do in beautie farre surpasse them.

WIND-FLOURES

The stocke or kindred of the *Anemones* or Winde-floures, especially in their varieties of colours, are without number,

Anemone

or at the least not sufficiently knowne unto any one that hath written of plants. My selfe have in my garden twelve different sorts: and yet I do heare of divers more differing very notably from any of these: every new yeare bringing with it new and strange kindes; and every country his peculiar plants of this sort which are sent unto us from far countries, in hope to receive from us such as our country yeeldeth.

The first kinde of *Anemone* or Winde-floure hath small leaves very much snipt or jagged almost like unto Camomile, or Adonis floure: among which riseth up a stalke bare or naked almost unto the top; at which place is set two or three leaves like the other: and at the top of the stalke commeth forth a faire and beautifull floure compact of seven leaves, and sometimes eight, of a violet colour tending to purple. It is impossible to describe the colour in his full perfection, considering the variable mixtures. The root is tuberous or knobby, and very brittle.

9

The second kind of *Anemone* hath leaves like to the precedent, insomuch that it is hard to distinguish the one from the other but by the floures onely: for those of this plant are of a most bright and faire skarlet colour, and as double as the Marigold; and the other not so.

The great *Anemone* hath double floures, usually called the *Anemone* of Chalcedon (which is a city in Bithynia) and great broad leaves deepely cut in the edges, not unlike to those of the field Crow-foot, of an overworne greene colour: amongst which riseth up a naked bare stalke almost unto the top, where there stand two or three leaves in shape like the others, but lesser; sometimes changed into reddish stripes, confusedly mixed here and there in the said leaves. On the top of the stalke standeth a most gallant floure very double, of a perfect red colour, the which is sometimes striped amongst the red with a little line or two of yellow in the middle; from which middle commeth forth many blackish thrums.

They floure from the beginning of Januarie to the end of April, at what time the flours do fade, and the seed flieth away with the wind, if there be any seed at all; the which I could never as yet observe.

Anemone, or Wind-floure, is so called, for the floure doth never open it selfe but when the wind doth blow, as *Pliny* writeth.

WALL-FLOURES, OR YELLOW STOCKE-GILLOFLOURES

The stalks of the Wal-floure are ful of greene branches, the leaves are long, narrow, smooth, slippery, of a blackish green color, and lesser than the leaves of stocke Gillofloures. The floures are small, yellow, very sweet of smell, and made of foure little leaves; which being past, their succeed long slender cods, in which is contained flat reddish seed. The whole plant is shrubby, of a wooddy substance, and can easily endure the cold of Winter.

The double Wal-floure hath long leaves greene and smooth, set upon stiffe branches, of a wooddy substance: whereupon doe grow most pleasant sweet yellow flours very double; which plant is so well knowne to all, that it shall be needlesse to spend much time about the description.

Of this double kinde we have another sort that bringeth his floures open all at once, whereas the other doth floure by degrees, by meanes whereof it is long in flouring.

The first groweth upon bricke and stone walls, in the corners of churches every where, as also among rubbish and other stony places. The double Wall-floure groweth in most gardens of England.

They floure for the most part all the yeere long, but especially in Winter, whereupon the people in Cheshire do call them Winter-Gillofloures.

The Wall-floure is called in Latine, *Viola lutea*, and *Leucoium luteum*: in English, Wall-Gillofloure, Wall-floure, yellow stocke Gillofloure, and Winter-Gillofloure.

The leaves stamped with a little bay salt, and bound about the wrests of the hands, take away the shaking fits of the Ague.

Water Crow-foot

Water Crow-foot hath slender branches trailing far abroad, whereupon grow leaves under the water, most finely cut and jagged: those above the water are somwhat round, in forme not unlike the smal tender leaves of the Mallow, but lesser: among which doe grow the floures, small, and white of colour, made of fine little leaves, with some yellownesse in the middle like the floures of the strawberry, and of a sweet smell. The roots be very small hairy strings.

Water Crowfoot growes by ditches and shallow springs, and in other moist and plashy places.

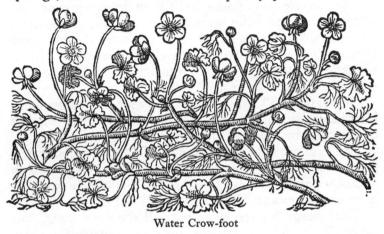

Water Crow-foot

CUCKOW PINT, OR WAKE-ROBIN

Arum or Cockow pint hath great, large, smooth, shining, sharpe pointed leaves, bespotted here and there with blackish spots, mixed with some blewnesse: among which riseth up a stalke, nine inches long, bespeckled in many places with certaine purple spots. It beareth also a certaine long hose or hood, in proportion like the eare of an hare: in the middle of which hood commeth forth a pestle or clapper of a darke murry or pale purple colour: which being past, there succeedeth in place thereof a bunch or cluster of berries in manner of a bunch of grapes, greene at the first, but after they be ripe of a yellowish red like corall, and full of juyce, wherein lie hid one or two little hard seeds. The root is tuberous, of the bignesse of a large Olive, white and succulent, with some threddy additaments annexed thereto.

Cockow pint groweth in woods neere unto ditches under hedges, every where in shadowie places. The leaves appeare presently after Winter: the pestell sheweth it selfe out of his huske or sheath in June, whilest the leaves are in withering: and when they are gone, the bunch or cluster of berries becommeth ripe, which is in July and August. The common Cuckow pint is called in Latine, *Arum*: in English, Cuckow pint, and Cuckow pintle, wake-Robin, Priests pintle, Aron, Calfes foot, and Rampe; and of some Starchwort.

Beares after they have lien in their dens forty daies without any manner of sustenance, but what they get with licking and sucking their owne feet, doe as soone as they come forth eat the herbe Cuckow-pint, through the windie nature thereof the hungry gut is opened and made fit again to receive sustenance: for by abstaining from food so long a time, the gut is shrunke or drawne so close together, that in a manner it is quite shut up, as *Aristole*, *Ælianus*, *Plutarch*, *Pliny*, and others do write.

The most pure and white starch is made of the roots of Cuckow-pint; but most hurtfull to the hands of the Laundresse that hath the handling of it, for it choppeth, blistereth, and maketh the hands rough and rugged, and withall smarting.

FROGGE-BIT

There floteth or swimmeth upon the upper parts of the water a small plant, which we usually call Frog-bit, having little round leaves, thicke and full of juyce, very like to the leaves of wall Peniwort: the floures grow upon long

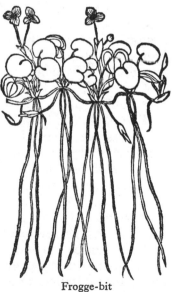

Frogge-bit

13

stems among the leaves, of a white colour, with a certain yellow thrum in the middle consisting of three leaves: in stead of roots it hath slender strings, which grow out of a short and small head, as it were, from whence the leaves spring, in the bottom of the water: from which head also come forth slopewise certain strings, by which growing forth it multiplieth it selfe.

It is found swimming or floting almost in every ditch, pond, poole, or standing water, in all the ditches about Saint George his fields, and in the ditches by the Thames side neere to Lambeth Marsh, where any that is disposed may see it. It flourisheth and floureth most part of all the yeare.

It is thought to be a kinde of Pond-weed (or rather of Water Lillie).

Little Daisies

The Daisie bringeth forth many leaves from a threddy root, smooth, fat, long, and somwhat round withall, very sleightly indented about the edges, for the most part lying upon the ground: among which rise up the floures, everie one with his owne slender stem, almost like those of Camomill, but lesser, of a perfect white colour, and very double.

The double red Daisie is like unto the precedent in everie respect, saving in the color of the floures; for this plant bringeth forth floures of a red colour; and the other white as aforesaid. The double Daisies are planted in gardens: the others grow wilde everywhere.

The Daisie is called of some, *Herba Margarita*, or Margarites herb: in French, *Marguerites*: In English, Daisies, and Bruisewort.

The Daisies do mitigate all kinde of paines, but especially in the joints, and gout, if they be stamped with new butter unsalted, and applied upon the pained place:

but they worke more effectually if Mallowes be added thereto.

The juice of the leaves and roots snift up into the nosthrils, purgeth the head mightily, and helpeth the megrim. The same given to little dogs with milke, keepeth them from growing great.

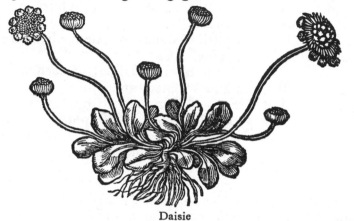

Daisie

The leaves stamped take away bruises and swellings proceeding of some stroke, if they be stamped and laid thereon; whereupon it was called in old time Bruisewort. The juice put into the eies cleareth them, and taketh away the watering of them. The decoction of the field Daisie (which is the best for physicks use) made in water and drunke, is good against agues.

GROUND-IVY, OR ALE-HOOFE

Ground Ivy is a low or base herbe; it creepeth and spreads upon the ground hither and thither all about, with many stalkes of an uncertaine length, slender, and like those of the Vine: whereupon grow leaves something broad and round: amongst which come forth the floures gaping like little hoods, not unlike to those of Germander, of a purplish blew colour: the whole plant is of a strong smell and bitter taste.

It is found as well in tilled as in untilled places, but most commonly in obscure and darke places, upon banks under hedges, and by the sides of houses.

It remaineth greene not onely in Summer, but also in Winter at any time of the yeare: it floureth from Aprill till Summer be far spent.

It is called in English Ground-Ivy, Ale-hoofe, Gill go by ground, Tune-hoof, and Cats-foot.

Ground-Ivy, Celandine, and Daisies, of each a like quantitie, stamped and strained, and a little sugar and rose water put thereto, and dropped with a feather ·into the eies, taketh away all manner of inflammation, spots, webs, itch, smarting, or any griefe whatsoever in the eyes, yea although the sight were nigh hand gone: it is proved to be the best medicine in the world.

The herbes stamped as aforesaid, and mixed with a little ale and honey, and strained, take away the pinne and web, or any griefe out of the eyes of horse or cow, or any other beast, being squirted into the same with a syringe, or I might have said the liquor injected into the eies with a syringe. But I list not to be over eloquent among Gentlewomen, to whom especially my Workes are most necessarie.

The women of our Northerne parts, especially about Wales and Cheshire, do turne the herbe Ale-hoof into their Ale; but the reason thereof I know not: notwithstanding without all controversie it is most singular against the griefes aforesaid; being tunned up in ale and drunke, it also purgeth the head from rheumaticke humors flowing from the braine.

GROUNDSELL

The stalke of Groundsell is round, chamfered and divided

into many branches. The leaves be green, long, and cut in the edges almost like those of Succorie, but lesse, like in a manner to the leaves of Rocket. The floures be yellow, and turn to down, which is carried away with the wind. The root is full of strings and threds.

These herbs are very common throughout England and do grow almost every where. They flourish almost every moneth of the yeare.

Groundsel is called in Latine *Senecio*, because it waxeth old quickly.

The leaves of Groundsel boiled in wine or water, and drunke, heale the paine and ach of the stomacke that proceeds of Choler. Stamped and strained into milke and drunke, they helpe the red gums and frets in Children.

Dioscorides saith, That with the fine pouder of Frankincense it healeth wounds in the sinues. The like operation hath the downe of the floures mixed with vineger.

Boiled in ale with a little hony and vineger, it provoketh vomit, especially if you adde thereto a few roots of *Asarabacca*.

Groundsell

Dandelion

The hearbe which is commonly called Dandelion doth send forth from the root long leaves deeply cut and gashed in the edges like those of wild Succorie,

17

Dandelion

but smoother: upon every stalke standeth a floure greater than that of Succorie, but double, & thicke set together, of colour yellow, and sweet in smell, which is turned into a round downy blowbal that is carried away with the wind. The root is long, slender, and full of milky juice, when any part of it is broken, as is the Endive or Succorie, but bitterer in tast than Succorie.

They are found often in medowes neere unto water ditches, as also in gardens and high wayes much troden. They floure most times in the yeare, especially if the winter be not extreme cold.

LETTUCE

Garden Lettuce hath a long broad leafe, smooth, and of a light greene colour: the stalke is round, thicke set with leaves full of milky juice, bushed or branched at the top: whereupon do grow yellowish floures, which turne into downe that is carried away with the winde. The seed sticketh fast unto the cottony downe, and flieth away likewise, white of colour, and somewhat long: the root hath hanging on it many long tough strings, which being cut or broken, do yeeld forth in like manner as doth the stalke and leaves, a juice like to milke. And this is the true description of the naturall Lettuce, and not of the artificiall; for by manuring, transplanting, and having a regard to the Moone and other circumstances, the leaves of the artificiall Lettuce are oftentimes transformed into another shape: for either they are curled, or else so drawne together, as they seeme to be like a Cabbage or headed Colewort, and the leaves which be within and in the middest are something white, tending to a very light yellow.

Lettuce delighteth to grow in a mannured, fat, moist, and dunged ground: it must be sowen in faire weather in places where there is plenty of water and prospereth best if it be sowen very thin. It may well be sowen at any time of the yeare, but especially at every first Spring, and so soone as Winter is done, till Summer be well nigh spent.

Garden Lettuce is called in Latine, *Lactuca sativa*, of the milky juice which issueth forth of the wounded stalks and roots.

Lettuce cooleth the heat of the stomacke, called the heart-burning; and helpeth it when it is troubled with choler: it quencheth thirst, and causeth sleepe.

Lettuce maketh a pleasant sallad, being eaten raw with vineger, oile, and a little salt: but if it be boiled it is sooner digested, and nourisheth more.

It is served in these daies, and in these countries in the beginning of supper, and eaten first before any other meat: which also *Martiall* testifieth to be done in his time, marvelling why some did use it for a service at the end of supper, in these verses:

> Tell me why Lettuce, which our Grandsires last
>   did eate,
> Is now of late become to be the first of meat?

Notwithstanding it may now and then be eaten at both those times to the health of the body: for being taken before meat it doth many times stir up appetite: and eaten after supper it keepeth away drunkennesse which commeth by the wine; and that is by reason that it staieth the vapours from rising up into the head.

Cowslips

The first, which is called in English the field Cowslip, is as common as the rest, therefore I shal not need to spend much time about the description.

The second is likewise well knowne by the name of Oxlip, and differeth not from the other save that the

Field Cowslip

floures are not so thicke thrust together, and they are fairer, and not so many in number, and do not smell so pleasant as the other: of which kind we have one lately come into our gardens, whose floures are curled and wrinkled after a most strange maner, which our women have named Jack-an-apes on horsebacke.

Double Paigle, the English garden Cowslip with double yellow floures, is so commonly knowne that it needeth no description.

The fourth is likewise known by the name of double Cowslips, having but one floure within another, which maketh the same once double, where the other is many times double, called by *Pena, Geminata,* for the likenesse of the floures, which are brought forth as things against nature, or twinnes.

The fifth being the common whitish yellow field Primrose, needeth no description.

The sixth, which is our garden double Primrose, of all the rest is of greatest beauty, the description whereof I refer unto your owne consideration.

The seventh is also very well known, being a Primrose with greenish floures somwhat welted about the edges.

Cowslips and Primroses joy in moist and dankish places, but not altogether covered with water: they are found in woods and the borders of fields. They flourish from Aprill to the end of May, and some one or other of them do floure all Winterlong.

They are commonly called *Primula veris*, because they are the first among those plants that doe floure in the Spring, or because they floure with the first. The greater sort, called for the most part Oxlips or Paigles, are named of divers *Herba S. Petri*: in English, Oxlip, and Paigle.

A practitioner of London who was famous for curing the phrensie, after he had performed his cure by the due observation of physick, accustomed every yeare in the moneth of May to dyet his Patients after this manner: Take the leaves and floures of Primrose, boile them a little in fountaine water, and

Field Primrose

in some rose and Betony waters, adding thereto sugar, pepper, salt, and butter, which being strained, he gave them to drinke thereof first and last.

The roots of Primrose stamped and strained, and the juice sniffed into the nose with a quill or such like, purgeth the brain, and qualifieth the pain of the megrim.

An unguent made with the juice of Cowslips and oile of Linseed, cureth all scaldings or burnings with fire, water, or otherwise.

BEARES EARES, OR MOUNTAINE COWSLIPS

This beautifull and brave plant hath thicke, greene, and fat leaves, somewhat finely snipt about the edges, not altogether unlike those of Cowslips, but smoother, greener, and nothing rough or crumpled: among which riseth up a slender round stem a handfull high, bearing a tuft of floures at the top, of a faire yellow colour, not much unlike to the floures of Oxe-lips, but more open and consisting of one only leafe like Cotiledon: the root is very threddy, and like unto the Oxe-lip.

They grow naturally upon the Alpish and Helvetian mountaines: most of them do grow in our London Gardens.

Either the antient writers knew not these plants, or else the names of them were not by them or their successors diligently committed unto posterity. *Matthiolus* and other later writers have given names according to the similitude, or of the shape that they beare unto other plants: they that dwell about the Alpes doe call it by reason of the effects thereof; for the root is amongst them in great request for the strengthning of the head, that when they are on the tops of places that are high, giddinesse and the swimming of the braine may not afflict them: it is there called the Rocke-Rose, for that it groweth upon the rockes, and resembleth the brave colour of the rose.

Those that hunt in the Alps and high mountaines after Goats and bucks, do as highly esteeme hereof as of *Doronicum*, by reason of the singular effects that it hath, but (as I said before) one especially, even in that it preventeth the losse of their best joynts (I meane their

22

neckes) if they take the roots hereof before they ascend
the rocks or other high places.

Pasque floures

The first of these Pasque floures hath many small leaves
finely cut or jagged, like those of Carrots: among which
rise up naked stalkes, rough and hairie; whereupon doe
grow beautifull floures bell fashion, of a bright delaied
purple colour: in the bottome whereof groweth a tuft of
yellow thrums, and in the middle of the thrums it thrusteth forth a small purple pointell; when the whole floure is
past there succedeth an head or knob compact of many
gray hairy lockes, and in the sollid parts of the knobs
lieth the seed flat and hoary, every seed having his owne
small haire hanging at it. The root is thicke and knobby,
of a finger long, running right downe, and therefore not
unlike to those of the *Anemone*, which it doth in all other
parts very notably resemble, and whereof no doubt this
is a kinde.

The white Passe floure hath many fine jagged leaves,
closely couched or thrust together, which resemble an
Holy-water sprinckle, agreeing with the other in roots,
seeds, and shape of floures, saving that these are of a
white colour, wherein chiefly consisteth the difference.

The Passe-floure groweth in France in untoiled places:
in Germany they grow in rough and stony places, and
oftentimes on rockes.

Those with purple floures do grow very plentifully in
the pasture or close belonging to the parsonage house of a
small village six miles from Cambrige, called Hilder-
sham: the Parson's name that lived at the impression
hereof was Mr. *Fuller*, a very kind and loving man, and
willing to shew unto any man the said close, who desired
the same.

They floure for the most part about Easter, which

hath mooved mee to name it *Pasque-Floure*, or Easter floure: and often they doe floure againe in September. In Cambridge-shire where they grow, they are named Coventrie bels.

There is nothing extant in writing among Authors of any peculiar vertue, but they serve onely for the adorning of gardens and garlands, being floures of great beautie.

Sweet Saint Johns and Sweet Williams

Sweet Johns have round stalkes as have the Gillofloures, (whereof they are a kinde) a cubit high, whereupon doe grow long leaves broader than those of the Gillofloure, of a greene grassie colour: the floures grow at the top of the stalkes, very like unto Pinkes, of a perfect white colour.

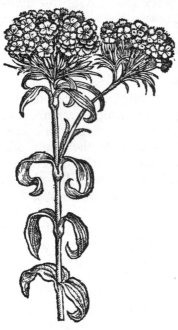

We have in our London Gardens a kinde hereof bearing most fine and pleasant white floures, spotted very confusedly with reddish spots, which setteth forth the beautie thereof; and hath bin taken of some (but not rightly) to be the plant called of the later Writers *Superba Austriaca*, or the Pride of Austria.

The great Sweet-William hath round joynted stalkes thicke and fat, some what reddish about the lower

Sweet-William

24

joynts, a cubit high, with long broad and ribbed leaves like as those of the Plantaine, of a greene grassie colour. The floures at the top of the stalkes are very like to the small Pinkes, many joyned together in one tuft or spoky umbell, of a deepe red colour.

These plants are kept and maintained in gardens more for to please the eye, than either the nose or belly. They are not used either in meat or medicine, but esteemed for their beauty to decke up gardens, the bosomes of the beautifull, garlands and crownes for pleasure.

STOCKE GILLO-FLOURES

The stalke of the great stocke Gillo-floure is two foot high or higher, round, and parted into divers branches. The leaves are long, white, soft, and having upon them as it were a downe like unto the leaves of willow, but softer: the floures consist of foure little leaves growing all along the upper part of the branches, of a white colour, exceeding sweet of smell: in their places come up long and narrow cods, in which is contained broad, flat, and round seed. The root is of a wooddy substance, as is the stalke also.

The purple stocke Gillo-floure is like the precedent in each respect, saving that the floures of this plant are of a pleasant purple colour, and the others white, which setteth forth the difference: of which kinde we have some that beare double floures which are of divers colours, greatly esteemed for the beautie of the floures, and pleasant sweet smell.

These kindes of Stocke Gillofloures do grow in most Gardens throughout England.

They floure in the beginning of the Spring, and continue flouring all the Summer long.

The Stocke Gillofloure is called in Latine, *Viola alba*: in Italian, *Viola bianca*: in Spanish, *Violetta blanquas*:

in English, Stocke Gillofloure, Garnsey Violet, and Castle Gillofloure.

They are not used in Physicke, except amongst certaine Empericks and Quacksalvers, about love and lust matters, which for modestie I omit.

White and blew Pipe Privet

The white Pipe groweth like an hedge tree or bushy shrub; from the root whereof arise many shoots which in short time grow to be equall with the old stocke, whereby in a little time it increaseth to infinit numbers, like the common English Prim or Privet, whereof doubtlesse it is a kinde, if wee consider every circumstance. The branches are covered with a rugged gray barke: the timber is white, with some pith or spongie matter in the middest like Elder, but lesse in quantitie. These little branches are garnished with small crumpled leaves of the shape and bignesse of the

White Pipe

Peare tree leaves, and very like in form: among which come forth the floures, growing in tufts, compact of four small leaves of a white colour, and of a pleasant sweet smell; but in my judgment they are too sweet, troubling and molesting the head in very strange manner. I once gathered the floures and layed them in my chamber window, which smelled more strongly after they had lien

together a few houres, with such an unacquainted savor
that they awaked me out of sleepe, so that I could not
rest till I had cast them out of my chamber. The floures
being vaded, the fruit follows, which is small, curled, and
as it were compact of many little folds, broad towards
the upper part, and narrow toward the stalke, and black
when it is ripe, wherein is contained a slender long seed.
The root hereof spreadeth it selfe abroad in the ground
after the manner of the roots of such shrubby trees.

The blew Pipe groweth likewise in manner of a small
hedge tree, with many shoots rising from the root like the
former, as our common Privet doth, whereof it is a kind.
The branches have a small quantity of pith in the middle
of the wood, and are covered with a darke blacke greenish
barke or rinde. The leaves are exceeding greene, and
crumpled or turned up like the brimmes of a hat, in shape
very like unto the leaves of the Poplar tree: among which
come the flours, of an exceeding faire blew colour, compact of many smal floures in the form of a bunch of grapes:
each floure is in shew like those of *Valeriana rubra
Dodonæi*, consisting of four parts like a little star, of an
exceeding sweet savour or smell, but not so strong as the
former. When these floures be gon there succeed flat cods,
and somwhat long, which being ripe are of a light colour,
with a thinne membrane or filme in the middest, wherein
are seeds almost foure square, narrow, and ruddy.

These trees grow not wild in England, but I have them
growing in my garden in very great plenty.

They floure in Aprill and May, but as yet they have not
borne any fruit in my garden, though in Italy and Spain
their fruit is ripe in September.

The later Physitians call the first *Syringa*, that is to say
a Pipe, because the stalkes and branches thereof when the
pith is taken out are hollow like a Pipe: it is also many
times syrnamed *Candida* or white, or *Syringa Candida
flore*, or Pipe with a white floure, because it should differ

from *Lillach*, which is somtimes named *Syringa cærulea* or blew Pipe.

Plum Tree

To write of Plums particularly would require a peculiar Volume, and yet the end not be attained unto, nor the stock or kindred perfectly known, neither to be distinguished apart: the numbers of the sorts or kinds are not known to any one Country, every clymat hath his own fruit, farre differing from that of other places: my selfe have sixty sorts in my garden, and all strange and rare: there be in other places many more common, and yet yearly commeth to our hands others not before known.

The Plum or Damson tree is of a mean bignesse, it is covered with a smooth barke: the branches are long, whereon do grow broad leaves more long than round, nicked in the edges: the floures are white; the plums do differ in colour, fashion, and bignesse, they all consist of pulp and skin, and also of kernell, which is shut up in a shell or stone. Some plums are of a blackish blew, of which some be longer, others rounder, others of the colour of yellow wax, divers of a crimson red, greater for the most part than the rest. There be also green plums, and withall very long, of a sweet and pleasant taste: moreover, the pulp or meat of some is drier, and easilier separated from the stone; of other-some it is moister, and cleaveth faster. Our common Damson is known to all, and therefore not to be stood upon.

The Mirobalan Plum tree groweth to the height of a great tree, charged with many great armes or boughes, which divide themselves into small twiggy branches, by means whereof it yeeldeth a goodly and pleasant shadow: the trunke or body is covered with a finer and thinner barke than any of the other Plum trees: the leaves do somewhat resemble those of the Cherrie tree, they are very tender, indented about the edges: the flours be white: the fruit is round, hanging upon long foot-stalks pleasant to behold, greene in the beginning, red when it is almost ripe, and beeing full ripe it glistereth like purple mixed with blacke: the flesh or meat is full of juice, pleasant in tast: the stone is small, or of a meane bignesse: the tree bringeth forth plenty of fruit every other yeare.

The Bullesse and the Sloe tree are wilde kindes of Plums, which do vary in their kind, even as the greater and manured Plums do. Of the Bullesse, some are greater and of better taste than others. Sloes are some of one taste, and some of others, more sharp; some greater, and others lesser; the which to distinguish with long descriptions were to small purpose, considering they be all and every of them knowne even unto the simplest: therefore this shall suffice for their severall descriptions.

The Plum trees grow in all knowne countries of the world: they require a loose ground, they also receive a difference from the regions where they grow, not only of the forme or fashion, but especially of the faculties.

The Plum trees are also many times graffed into trees of other kindes.

The wilde Plums grow in most hedges through England.

The common and garden Plum trees do bloome in April: the leaves come forth presently with them: the fruit is ripe in Summer, some sooner, some later.

Plummes that be ripe and new gathered from the tree, what sort soever they are of, do moisten and coole, and yeeld unto the body very little nourishment, and the same nothing good at all.

Dried Plums, commonly called Prunes, are wholesomer, and more pleasant to the stomack, they yeeld more nourishment and better.

CHERRY TREE

The English Cherry tree groweth to an high and great tree, the body whereof is of a mean bignesse, which is parted above into very many boughes, with a barke somewhat smooth, of a brown crimson colour, tough and pliable; the substance or timber is also brown in the

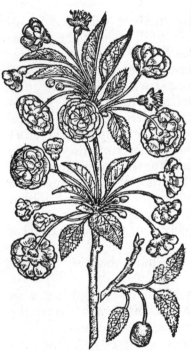

middle, and the outer part is somwhat white: the leaves be great, broad, long, set with veins or nerves, and sleightly nicked about the edges: the floures are white, of a mean bignes, consisting of five leaves, and having certain threds in the middle of the like colour. The Cherries be round, hanging upon long stems or footstalks, with a stone in the middest which is covered with a pulp or soft meat; the kernell thereof is not unpleasant to the taste, though somwhat bitter.

The late ripe Cherry tree groweth up like unto our wild English Cherry tree, with the like leaves, branches and floures, sav-

Double-floured Cherry

ing that they are somtimes once doubled: the fruit is small, round, and of a darke bloudy colour when they be ripe, which the French-men gather with their stalkes, and hang them up in their houses in bunches or handfulls against Winter, which the Physitions do give unto their patients in hot and burning fevers, being first steeped in a little warme water, that causeth them to swell and plumpe as full and fresh as when they did grow upon the tree.

The double floured Cherry-tree growes up like unto an hedge bush, but not so great nor high as any of the others; the leaves and branches differ not from the rest of the Cherry-trees. The floures hereof are exceeding double, as are the flours of Marigolds, but of a white colour, and smelling somewhat like the Hawthorne floures; after, which come seldome or never any fruit, although some Authors have said that it beareth sometimes fruit, which my selfe have not at any time seen; notwithstanding the tree hath growne in my Garden many yeeres, and that in an excellent good place by a bricke wall, where it hath the reflection of the South Sunne, fit for a tree that is not willing to beare fruit in our cold climat.

My selfe with divers others have sundry other sorts in our gardens, one called the Hart Cherry, the greater and the lesser; one of the great bignesse, and most pleasant in taste, which we call *Luke Wardes* Cherry, because he was the first that brought the same out of Italy; another we have called the Naples Cherry, because it was first brought into these parts from Naples: the fruit is very great, sharpe pointed, somewhat like a mans heart in shape, of a pleasant taste, and of a deepe blackish colour when it is ripe, as it were of the colour of dried bloud.

We have another that bringeth forth Cherries also very great, bigger than any Flanders Cherrie, of the colour of Jet, or burnished horne, and of a most pleasant taste, as witnesseth Mr. *Bull*, the Queenes Majesties Clocke-maker, who did taste of the fruit (the tree bearing onely one Cherry, which he did eate, but my selfe never tasted of it) at the impression hereof. We have also another, called the Agriot Cherry, of a reasonable good taste. Another we have with fruit of a dun colour, tending to a watchet. We have one of the dwarfe Cherries, that bringeth forth fruit as great as most of our Flanders Cherries, whereas the common sort hath very small Cherries, and those of an harsh taste. These and many sorts more we have in our London gardens, whereof to write particularly would greatly enlarge our volume, and to small purpose: therefore what hath beene said shall suffice.

The Cherrie-trees bloome in Aprill; some bring forth their fruit sooner; some later: the red Cherries be alwaies better than the blacke of their owne kinde.

Many excellent Tarts and other pleasant meats are made with Cherries, sugar, and other delicat spices.

PEARE TREE

To write of Peares and Apples in particular, would require a particular volume: the stocke or kindred of Peares are not to be numbred: every country hath his peculiar fruit: my selfe knowes one curious in graffing & planting of fruits, who hath in one piece of ground, at the point of three-score sundry sorts of Peares, and those exceeding good, not doubting but if his minde had been to seeke after multitudes, he might have gotten together the like number of those of worse kinds: besides the diversities of those that be wilde, experience sheweth sundry sorts: and therefore I thinke it not amisse to set downe one generall description for that, that might be said of many, which to describe apart, were to send an owle to Athens, or to number those things which are without number.

The Peare tree is for the most part higher than the Apple tree, having boughes not spread abroad, but growing up in height; the body is many times great: the timber or wood it selfe is very tractable or easie to be wrought upon, exceeding fit to make moulds or prints to be graven on, of colour tending to yellownesse: the leafe is somewhat broad, finely nicked in the edges, greene above, and somewhat whiter underneath: the floures are white: the Peares, that is to say, the fruit, are for the most part long, and in forme like a Top; but in greatnesse, colour, forme, and taste very much differing among themselves; they be also covered with skins or coats of sundry colours: the pulpe or meate differeth, as well in colour as taste: there is contained in them kernels, blacke when they be ripe: the root groweth straight downe with some branches running aslope.

The wilde Peare tree growes likewise great, upright, full of branches, for the most part Pyramides like, or of the fashion of a steeple, not spred abroad as is the Apple or Crab tree: the timber of the trunke or body of the tree is very firme and sollid, and likewise smooth, a wood very fit to make divers sorts of instruments of, as also the hafts of sundry tooles to worke withal; and likewise serveth to be cut into many kindes of moulds, not only such prints as these figures are made of, but also many sorts of pretty toies, for coifes, brest-plates, and such like, used among our English gentlewomen.

The tame Peare trees are planted in Orchards, as be the apple trees, and by grafting, though upon wilde stockes, come much variety of good and pleasant fruits.

The floures doe for the most part come forth in Aprill, the leaves afterwards: all peares are not ripe at one time: some be ripe in July, others in August, and divers in September and later.

Wine made of the juyce of Peares called in English, Perry, is soluble, purgeth those that are not accustomed to drinke thereof, especially when it is new; notwithstanding it is as wholesome a drinke being taken in small quantitie as wine; it comforteth and warmeth the stomacke, and causeth good digestion.

SPERAGE OR ASPARAGUS

The manured or garden Sperage, hath at his first rising out of the ground thicke tender shoots very soft and brittle, of the thicknesse of the greatest swans quill, in taste like the green bean, having at the top a certaine scaly soft bud, which in time groweth to a branch of the height of two cubits, divided into divers other smaller branches, wheron are set many little leaves like haires, more fine than the leaves of Dill: amongst which come forth small mossie yellowish floures which yeeld forth the fruit, green at the first, afterward as red as Corall, of the bignesse of a small pease; wherein is

Garden Sperage

contained grosse blackish seed exceeding hard, which is the cause that it lieth so long in the ground after his sowing, before it spring up: the roots are many thicke soft and spongie strings hanging downe from one head, and spred themselves all about, whereby it greatly increaseth.

Our garden Asparagus groweth wilde in Essex, in a medow neere to a mill, beyond a village called Thorp; and also at Singleton not far from Carby, and in the medowes neere Moulton in Lincolnshire. Likewise it growes in great plenty neere Harwich, at a place called Bandamar lading, and at North Moulton in Holland a part of Lincolnshire.

The bare naked tender shoots of Sperage spring up in Aprill, at what time they are eaten in sallads; they floure in June and July, the fruit is ripe in September.

It is named Asparagus, of the excellency, because *asparagi*, or the springs hereof are preferred before those of other plants whatsoever: for this Latine word *Asparagus* doth properly signifie the first spring or sprout of every plant, especially when it is tender, and before it do grow into an hard stalk, as are the buds, tendrels, or yong springs of wild Vine or hops, and such like.

The first sprouts or naked tender shoots hereof be oftentimes sodden in flesh broth and eaten; or boiled in faire water, and seasoned with oile, vineger, salt, and pepper, then are served up as a sallad: they are pleasant to the taste.

HORSE-TAILE OR SHAVE-GRASSE

Great Horse-taile riseth up with a round stalke hollow within like a reed, a cubit high, compact as it were of many small pieces one put into the end of another, somtimes of a reddish colour, very rough, and set at every joint with many stiffe Rush-like leaves, or rough bristles, which maketh the whole plant to resemble the taile of a horse, whereof it tooke his name.

Small and naked Shave-grasse, wherewith Fletchers and Combe-makers doe rub and polish their worke, riseth out of the ground like the first shoots of Asparagus, jointed or kneed by certain distances like the precedent, but altogether without such bristly leaves, yet exceeding rough and cutting: the root groweth aslope in the earth like those of the Couch-grasse.

Dodonæus sets forth another Horse-taile, which he called climing Horse-taile, or Horse-tail of Olympus. *Bellonius* writes in his Singularities, That it hath bin seen to be equall in height with the Plane tree.

Shave-grasse is not without cause named *Asprella*, of

his ruggednesse, which is not unknowne to women, who scoure their pewter and wodden things of the kitchen therewith: and therefore some of our huswives do call it Pewter-wort.

Dioscorides saith, that Horse-taile being stamped and laid to, doth perfectly cure wounds, yea although the sinues be cut in sunder, as *Galen* addeth.

The herb drunke either with water or wine, is an excellent remedy against bleeding at the nose. Horse-taile with his roots boiled in wine is very profitable for difficultie of breathing.

Colts-foot, or Horse-foot

Tussilago or Fole-foot hath many white and long creeping roots, somewhat fat; from which rise up naked stalkes (in

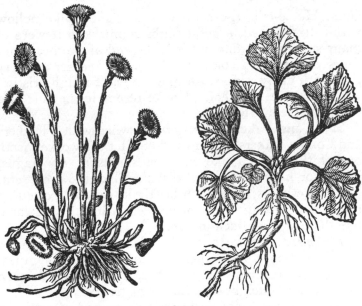

Colts-foot

the beginning of March and Aprill) about a spanne long, bearing at the top yellow floures, which change into downe and are caried away with the winde: when the stalke and seed is perished, there appeare springing of out the earth many broad leaves, greene above, and next the ground of a white hoarie or grayish colour, fashioned like an Horse foot; for which cause it was called Fole-foot, and Horse-hoofe: seldome or never shall you find leaves and floures at once, but the flours are past before the leaves come out of the ground; as may appeare by the first picture, which setteth forth the naked stalkes and floures; and by the second, which pourtraiteth the leaves only.

A decoction made of the greene leaves and roots, or else a syrrup thereof, is good for the cough. The fume of the dried leaves taken through a funnell or tunnell, burned upon coles, effectually helpeth those that are troubled with the shortnesse of breath, and fetch their winde thicke and often. Being taken in manner as they take Tobaco, it mightily prevaileth against the diseases aforesaid.

HEARTS-EASE, OR PANSIES

The Hearts-ease or Pansie hath many round leaves at the first comming up; afterward they grow somewhat longer, sleightly cut about the edges, trailing or creeping upon the ground: the stalks are weake and tender, whereupon grow floures in form & figure like the Violet, and for the most part of the same bignesse, of three sundry colours, whereof it took the syrname *Tricolor*, that is to say, purple, yellow, and white or blew; by reason of the beauty and braverie of which colours they are very pleasing to the eye, for smel they have little or none at all.

There is found in sundry places of England a wilde kinde hereof, having floures of a feint yellow colour, without mixture of any other colour, yet having a deeper

37

yellow spot in the lowest leafe with foure or five blackish purple lines, wherein it differeth from the other wilde kinde: and this hath been taken of some young Herbarists to be the yellow Violet.

The Hearts-ease groweth in fields in many places, and in gardens also, and that oftentimes of it selfe: it is more gallant and beautifull than any of the wilde ones.

CROWNE IMPERIALL

Crowne Imperiall

The Crowne Imperiall hath for his root a thicke firme and solid bulbe, covered with a yellowish filme or skinne, from the which riseth up a great thicke fat stalke two cubits high, in the bare and naked part of a darke overworne dusky purple colour. The leaves grow confusedly about the stalke like those of the white Lilly, but narrower: the floures grow at the top of the stalke, incompassing it round, in forme of an Imperiall Crowne, (whereof it tooke his name) hanging their heads downward as it were bels; in colour it is yellowish; or to give you the true colour, which by words otherwise cannot be expressed, if you lay sap berries in steep in faire water for the space of two houres, and mix a little saffron in that infusion, and lay it upon paper, it sheweth the perfect colour to limne or illumine the floure withall. The back side of the said floure is streaked with purplish lines, which doth greatly set forth the beauty therof. In the bottom of each of these bels there is placed six drops of most cleare shining sweet water, in taste like sugar resembling in shew faire orient pearles; the which drops if you take away, there do immediatly appeare the like: notwithstanding if they may be suffered to stand still in the floure according to his own nature, they will never fall away, no not if you strike the plant untill it be broken. Among these drops there standeth out a certain pestel, as also sundry small chives tipped with small pendants like those of the Lilly: above the whole floures there groweth a tuft of green leaves like those upon the stalke, but smaller. After the floures be faded, there follow cods or seed-vessels six square, wherein is contained flat seeds tough & limmer, of the colour of Mace: the whole plant, as wel roots as floures do savor or smell very like a fox. As the plant groweth old, so doth it wax rich, bringing forth a Crowne of floures amongst the uppermost green leaves, which some make a second kinde, although in truth they are but one and the selfe same, which in time is thought to grow to a triple crowne, which hapneth by the age of the root, and fertilitie of the soile.

This plant hath been brought from Constantinople amongst other bulbous roots, and made denizons in our London gardens, whereof I have great plenty.

It floureth in Aprill, and sometimes in March, when as the weather is warme and pleasant.

GREAT CELANDINE OR SWALLOW-WORT

The great Celandine hath a tender brittle stalke, round, hairy, and full of branches, set with leaves not unlike to those of Columbine, but tenderer, and deeper cut or jagged, of a grayish green under, and greene on the other side tending to blewnesse: the floures grow at the top of the stalks, of a gold yellow colour, in shape like those of the Wal-floure: after which come long cods full of bleak or pale seeds: the whole plant is of a strong unpleasant smell, and yeeldeth a thicke juice of a milky substance, of the colour of Saffron: the root is thicke and knobby, with some threds anexed thereto, which beeing broken or bruised, yeeldeth a sap or juice of the colour of gold.

It groweth in untilled places by common way sides, among briers and brambles, about old wals, and in the shade rather than in the Sun.

It is greene all the yeare: it floureth from Aprill to a good part of Summer: the cods are perfected in the mean time.

It is called in Latine *Chelidonium majus,* and *Hirundinarium major*: in English, Celandine, Swallowwort, and Tetter-wort.

It is called Celandine not because it first springeth at the comming in of Swallowes, or dieth when they go away, (for as we have said, it may be found all the yere) but because some hold opinion, that with this herb the dams restore sight to their yong ones when they

Great Celandine

cannot see. Which things are vain and false; for *Cornelius Celsus, lib.* 6. witnesseth, That when the sight of the eies of divers yong birds is put forth by some outward means, it will after a time be restored of it selfe, and soonest of all the sight of the Swallow: whereupon (as the same Author saith) the tale grew, how thorow an herb the dams restore that thing which healeth of it selfe. The very same doth *Aristotle* alledge, *lib.* 6. *de Animal.* The eies of Swallowes (saith he) that are not fledge, if a man do pricke them out, do afterwards grow againe and perfectly recover their sight.

The juice of the herbe is good to sharpen the sight, for it clenseth and consumeth away slimie things that cleave about the ball of the eye, and hinder the sight, and especially being boiled with hony in a brasen vessell.

The root being chewed is reported to be good against the toothache.

The root cut into small pieces is good to be given unto Hauks against sundry diseases, wherunto they are subject.

Wood Sorrell, or Stubwort

Oxys Pliniana, or *Trifolium acetosum*, being a kinde of three leafed grasse, is a low and base herbe without stalke; the leaves immediately rising from the root upon short stemmes at their first comming forth folded together, but afterward they do spred abroad, and are of a faire light greene colour, in number three, like the rest of the Trefoiles, but that each leafe, hath a deep cleft or rift in the middle: among these leaves come up small and weake tender stems, such as the leaves do grow upon, which beare small starre-like floures of a white colour, with some brightnes of carnation dasht over the same: the floure consisteth of five small leaves; after which come little round knaps or huskes full of yellowish seed.

Wood Sorrell

These plants grow in

woods and under bushes, in sandie and shadowie places in every countrey.

Wood Sorrell or Cuckow Sorrell is called in Latine *Trifolium acetosum*: the Apothecaries and Herbarists call it *Alleluya*, and *Panis Cuculi*, or Cuckowes meate, because either the Cuckow feedeth thereon or by reason when it springeth forth and floureth the Cuckow singeth most, at which time also *Alleluya* was wont to be sung in Churches.

Sorrell du Bois or Wood Sorrell stamped and used for greene sauce, is good for them that have sicke and feeble stomackes; for it strengthneth the stomacke, procureth appetite, and of all Sorrell sauces is the best, not onely in vertue, but also in the pleasantnesse of his taste.

D o c k e

The great water-dock hath very long and great leaves, stiffe and hard, not unlike to the garden Patience, but much longer. The stalke riseth up to a great height, oftentimes to the height of five foot or more. The floure groweth at the top of the stalk in spoky tufts, brown of colour. The seed is contained in chaffie husks three square, of a shining pale colour. The root is very great, thick, brown without and yellowish within.

The smal water-Dock hath short narrow leaves set upon a stiffe stalke. The floures grow from the middle of the stalke upward in spoky rundles, set in spaces by certain distances round about the stalk, as are the floures of Horehound: which Docke is of all the kinds most common, and of least use, and takes no pleasure or delight in any one soile or dwelling place, but is found almost every where, as well upon the land as in waterie places, but especially in gardens among good and wholsome pot-herbs, being there better knowne, than welcome or desired: wherefor I intend not to spend farther time about his description.

The garden Patience hath very strong stalks furrowed or chamfered, of eight or nine foot high when it groweth in fertile ground, set about with great large leaves like to those of the water-Docke, having alongst the stalkes toward the top floures of a light purple colour declining to brownnesse. The seed is three square, contained in thin chaffie husks like those of the common Docke. The root is very great, browne without and yellow within, in colour and taste like the true Rubarb.

Bloudwort is best knowne unto all of the stocke or kindred of Dockes: it hath long thin leaves sometimes red in every part thereof, and often striped here & there with lines and strakes of a darke red colour: among which rise up stiffe brittle stalkes of the same colour: on the top whereof come forth such floures and seed as the common wild Docke hath. The root is likewise red, or of a bloudy colour.

The Monks Rubarb is called Patience, which word is borrowed of the French, who call this herb *Patience*: of some Monks Rubarb, because as it seemes some Monke or other hath used the root hereof in stead of Rubarb.

Bloudwort or bloudy Patience is called of some, *Sanguis Draconis*, of the bloudy colour wherewith the whole plant is possest: it is of pot-herbs the chiefe or principall, having the propertie of the bastard Rubarb, but of lesse force in his purging qualitie.

Monks Rubarb or Patience is an excellent wholsome pot-herb, for being put into the pottage in some reasonable quantitie, it helps the jaundice, and such like diseases proceeding of cold causes.

If you take the roots of Monks Rubarb and red Madder of each halfe a pound, Sena foure ounces, Anise seed and Licorice of each two ounces, Scabious and Agrimonie of each one handfull; slice the roots of the Rubarb, bruise the Anise seed and Licorice, breake the herbs with your hands and put them into a stone pot called a

43

stean, with foure gallons of strong ale, to steep or infuse the space of three daies, and then drinke this liquor as your ordinary drink for three weeks together at the least, though the longer you take it, so much the better; providing in a readinesse another stean so prepared, that you may have one under another, being alwaies carefull to keep a good diet: it purifieth the bloud and makes yong wenches look faire and cherry-like.

There have not bin any other faculties attributed to this plant, either of the antient or later writers, but generally of all it hath bin referred to the other Docks or Monks Rubard: of which number I assure my selfe this is the best, and doth approch neerest unto the true Rubarb. Other distinctions and differences I leave to the learned Physitions of our London colledge, who are very well able to search this matter, as a thing far above my reach, being no Graduat, but a Country Scholler, as the whole frame of this historie doth well declare: but I hope my good meaning will be well taken, considering I do my best: and I doubt not but some of greater learning wil perfect that which I have begun according to my small skill, especially the ice being broken to him, and the wood rough-hewn to his hand. Notwithstanding I thinke it good to say thus much more in my own defence, That although there be many wants and defects in mee, that were requisite to performe such a worke; yet may my long experience by chance happen upon some one thing or other that may do the Learned good.

D U C K S   M E A T

Ducks meat is as it were a certain green mosse, with very little round leaves of the bignes of Lentils: out of the midst whereof on the nether side grow downe very fine threds like haires, which are to them in stead of roots: it hath neither stalke, floure, nor fruit.

It is found in ponds, lakes, city ditches, & other standing waters every-where.

The time of Ducks meat is known to all.

Duckes meat is called Ducks herb, because Ducks do feed thereon; where-upon also it is called Ducks meat.

Ducks meat mingled with fine wheaten floure, and applied, prevaileth much against hot Swell-ings.

Cuckow-floures

The first of the Cuckow flours hath leaves at his springing up somwhat

Ducks meat

round, & those that spring afterward grow jagged like the leaves of Greek Valerian; among which riseth up a stalk a foot long, set with the like leaves, but smaller and more jagged, resembling those of Rocket. The floures grow at the top in small bundles, white of colour, hollow in the middle, resembling the white sweet-John: after which come small chaffie huskes or seed-vessels, wherein the seed is contained.

These floure for the most part in Aprill and May, when the Cuckow begins to sing her pleasant notes with-out stammering.

They are commonly called in Latine *Flos Cuculi*; and also some call them *Nasturtium aquaticum minus*, or lesser water-Cresse: of some, *Cardamine*: in English, Cuckow

45

flours: in Norfolk, Canturbury bels: at the Namptwich in Cheshire my native country, Lady-smockes.

BROOME

Broom is a bush or shrubby plant, it hath stalks or rather

wooddy branches, from which do spring slender twigs, cornered, green, tough, and that be easily bowed, many times divided into smal branches; about which do grow little leaves of an obscure green colour, & brave yellow floures, and at the length flat cods, which beeing ripe are black, as are those of the common Vetch, in which doe lie flat seeds, hard, something brownish, and lesser than Lentils.

The Spanish Broome hath likewise wooddy stems, from whence grow up slender pliant twigs, which be bare and naked without leaves, or at the least having but few small leaves, set here and there far

Spanish Broome

distant one from another, with yellow floures not much unlike the floures of common Broome, but greater.

Small leafed or thin leafed Broome hath many tough pliant shoots rising out of the ground, which grow into hard and tough stalks, which are divided into divers twiggy branches whereon doe grow very small thin leaves, of a whitish colour; whereupon some have called it *Genista alba*, white Broome: the floures grow at the top of the stalkes, in shape like those of the common Broom

46

but of a white colour, wherein it specially differeth from the other Broomes.

The common Broome groweth almost every where in dry pastures and low woods. Spanish Broome groweth in divers kingdomes of Spaine and Italy; we have it in our London gardens. The White Broome groweth likewise in Spaine and other hot regions; it is a stranger in England; of this *Titus Calphurnius* makes mention in his second Eclog of his Bucolicks, writing thus:

See Father, how the Kine stretch out their
    tender side
Under the hairy Broome, that growes in field
    so wide.

Broome floureth in the end of Aprill or May, and then the young buds of the floures are to bee gathered and laid in pickle or salt, which afterwards being washed or boyled, are used for sallads, as Capers be, and be eaten with no lesse delight. The Spanish Broome doth floure sooner, and is longer in flouring.

There is made of the ashes of the stalkes and branches dryed and burnt, a lie with thin white wine, as Rhenish wine, which is highly commended of divers for the greene sickenesse and dropsie; but withall it doth by reason of his sharpe quality many times hurt and fret the intrailes.

The young buds or little floures preserved in pickle, and eaten as a sallad, stirre up an appetite to meate.

That worthy Prince of famous memory *Henry* 8. King of England, was wont to drinke the distilled water of Broome floures, against surfets and diseases thereof arising.

Sir *Thomas Fitzherbert* Knight, was wont to cure the blacke jaundice with this drink onely. Take as many handfulls (as you thinke good) of the dried leaves of Broome gathered and brayed to pouder in the moneth of May, then take unto each handfull of the dried leaves, one spoonful and a halfe of the seed of Broome brayed into pouder: mingle these together, and let the sicke drinke thereof each day a quantity, first and last, untill he finde some ease. The medicine must be continued and so long used, untill it be quite extinguished: for it is a disease not very suddenly cured, but must by little and little be dealt withall.

Furze, Gorsse, Whin, or prickley Broome

There be divers sorts of prickely Broome, called in our English tongue by sundry names, according to the speech of the countrey people where they doe grow: in some places, Furzes; in others, Whins, Gorsse, and of some, prickly Broome.

The Furze bush is a plant altogether a Thorne, fully armed with most sharpe prickles, without any leaves at all except in the Spring, and those very few and little, and quickly falling away: it is a bushy shrub, often rising up with many wooddy branches to the height of foure or five cubits or higher, according to the nature and soile where they grow: the greatest and highest that I did ever see do grow about Excester in the West parts of England, where the great stalks are dearely bought for the better sort of people, and the small thorny spraies for the poorer sort. From these thorny branches grow little floures like those of Broome, and of a yellow colour, which in hot regions under the extreme heat of the Sunne are of a very perfect red colour: in the colder countries of the East, as Danzicke, Brunswicke, and Poland, there is not any branch hereof growing, except some few plants and seeds which my selfe have sent to Elbing otherwise called Meluin, where they are most curiously kept in their fairest gardens, as also our common Broome, the which I have sent thither likewise, being first desired by divers earnest letters.

We have in our barren grounds of the North part of England another sort of Furze, bringing forth the like prickly thornes that the other have: the onely difference consisteth in the colour of the floures; for the others bring forth yellow floures and those of this plant are as white as snow.

Petty Whin (growing upon Hampstead heath neere London, and in divers other barren grounds, where, in manner nothing else will grow) hath many weake and flexible branches of a wooddy substance: whereon doe grow little leaves like those of Tyme: among which are set in number infinite most sharpe prickles, hurting like needles, whereof it tooke his name. The floures grow on the tops of the branches like those of Broome, and of a pale yellow colour.

WILLOW TREE

The common Willow is an high tree, with a body of a meane thicknesse, and riseth up as high as other trees doe if it be not topped in the beginning, soone after it is planted; the barke thereof is smooth, tough, and flexible: the wood is white, tough, and hard to be broken: the leaves are long, lesser and narrower than those of the Peach tree, somewhat greene on the upper side and slipperie, and on the nether side softer and whiter: the boughes be covered either with a purple, or else with a white barke: the catkins which grow on the toppes of the branches come first of all forth, being long and mossie, and quickly turne into white and soft downe, that is carried away with the winde.

The Oziar or Water Willow bringeth forth of the head, which standeth somewhat out, slender wands or twigs, with a reddish or greene barke, good to make baskets and such like workes of: it is planted by the twigs or cods being thrust into the earth, the upper part whereof

Common Willow

when they are growne up, is cut off, so that which is called the head increaseth under them, from whence the slender twigs doe grow, which being oftentimes cut, the head waxeth greater: many times also the long rods or wands of the higher Withy trees be lopped off and thrust into the ground for plants, but deeper, and above mans height: of which do grow great rods, profitable for many things, and commonly for bands, wherewith tubs and casks are bound.

The Sallow tree or Goats Willow, groweth to a tree of a meane bignesse: the trunke or body is soft and hollow timber, covered with a whitish rough barke: the branches are set with leaves somewhat rough, greene above, and hoarie underneath: among which come forth round catkins, or aglets that turne into downe, which is carried away with the winde.

The Rose Willow groweth up likewise to the height and bignesse of a shrubby tree; the body whereof is covered with a scabby rough barke: the branches are many, whereupon do grow very many twigs of a reddish colour, garnished with small long leaves, somewhat whitish: amongst which come forth little floures, or rather a multiplication of leaves, joyned together in forme of a Rose, of a greenish white colour, which doe not only make a gallant shew, but also yeeld a most cooling aire in the heat of Summer, being set up in houses, for the decking of the same.

These Willowes grow in divers places of England: the Rose-Willow groweth plentifully in Cambridge shire, by the rivers and ditches there in Cambridge towne they grow abundantly about the places called Paradise and Hell-mouth, in the way from Cambridge to Grand-chester.

The greene boughes with the leaves may very well be brought into chambers and set about the beds of those that be sicke of fevers, for they doe mightily coole the heate of the aire, which thing is a wonderfull refreshing to the sicke Patients.

The barke hath like vertues: *Dioscorides* writeth, That this being burnt to ashes, and steeped in vineger, takes away cornes and other like risings in the feet and toes.

LARCH TREE

The Larch is a tree of no small height, with a body growing straight up: the barke whereof in the neither part beneath the boughes is thicke, rugged and full of chinkes; which being cut in sunder is red within, and in the other part above smooth, slipperie, something white without: it bringeth forth many boughes divided into other lesser branches, which be tough and pliable. The leaves are small and cut into many jags, growing in clusters thicke together like tassels, which fall away at the approch of Winter: the floures, or rather the first shewes of the cones or fruit be round, and grow out of the tenderest boughes, being at the length of a brave red purple colour: the cones be small, and like almost in bignesse to those of the Cypresse tree, but longer, and made up of a multitude of thin scales like leaves: under which lie small seeds, having a thin velme growing on them very like to the wings of Bees and wasps: the substance of the wood is very hard of colour, especially that in the midst somewhat red, and very profitable for workes of long continuance.

It is not true that the wood of the Larch tree cannot be set on fire, as *Vitruvius* reporteth of the castle made of Larch wood, which *Cæsar* besieged, for it burneth in chimneies, and is turned into coles, which are very profitable for Smithes.

There is also gathered of the Larch tree a liquid Rosin, very like in colour and substance to the whiter hony, as that of Athens or of Spaine, which notwithstanding issueth not forth of it selfe, but runneth out of the stocke of the tree, when it hath beene bored even to the heart with a great and long auger and wimble.

Of all the Cone trees onely the Larch tree is found to be without leaves in the Winter: in the Spring grow fresh leaves out of the same knobs, from which the former did fall. The cones are to be gathered before Winter, so soon as the leaves are gone: but after the scales are loosed and opened, the seeds drop away: the Rosine must be gathered in the Summer moneths.

White Thorne or Hawthorne Tree

The white Thorn is a great shrub growing often to the height of a peare tree, the trunk or body is great, the boughes and branches hard and woody, set with long sharp thorns: the leaves be broad, cut with deep gashes into divers sections, smooth, and of a glistering green colour: the floures grow upon spoky rundles, of a pleasant sweet smell, somtimes white, and often dasht over with a light wash of purple, which have moved some to thinke a difference in the plants: after which come the fruit, being round berries, green at the first, and red when they be ripe; wherein is found a soft sweet pulpe and certaine whitish seed: the root growes deepe in the ground, of a hard wooddy substance.

The Hawthorne groweth in woods and in hedges neere unto highwaies almost every where.

The Hawthorne floures in May, whereupon many do call the tree it selfe the May-bush, as a chiefe token of the comming in of May: the leaves come forth a little sooner: the fruit is ripe in the beginning of September, and is a food for birds in Winter.

Lilly in the Valley, or May Lilly

Lilly of the Valley

The Convall Lilly, or Lilly of the Vally, hath many leaves like the smallest leaves of Water Plantaine; among which riseth up a naked stalke halfe a foot high, garnished with many white floures like little bels, with blunt and turned edges, of a strong savour, yet pleasant enough; which being past, there come small red berries, much like the berries of *Asparagus*, wherein the seed is contained. The root is small and slender creeping far abroad in the ground.

The second kinde of May Lillies is like the former in every respect; and herein varieth or differeth, in that this kinde hath reddish floures, and is thought to have the sweeter smell.

The first groweth on Hampsted heath, foure miles from London, in great abundance: neere to Lee in Essex, and upon Bushie heath, thirteene miles from London, and many other places.

The other kinde with the red floure is a stranger in England: howbeit I have the same growing in my garden.

They floure in May, and their fruit is ripe in September.

The Latines have named it *Lilium Convallium*: in French, *Muguet*: yet there is likewise another herbe which they call *Muguet*, commonly named in English, Woodroof. It is called in English, Lilly of the Valley, or the Convall Lillie, and May Lillies, and in some places Liriconfancie.

The floures of the Valley Lillie distilled with wine, and drunke the quantitie of a spoonefull, restore speech unto those that have the dumb palsie and that are falne into the Apoplexie, and are good against the gout, and comfort the heart.

The water aforesaid doth strengthen the memory that is weakened and diminished; it helpeth also the inflammations of the eies, being dropped thereinto.

The floures of May Lillies put into a glasse, and set in a hill of ants, close stopped for the space of a moneth, and then taken out, therein you shall finde a liquor that appeaseth the paine and griefe of the gout, being outwardly applied; which is commended to be most excellent.

ENGLISH JACINTH, OR HARE-BELLS

The blew Hare-bells or English Jacinth is very common throughout all England. It hath long narrow leaves leaning towards the ground, among the which spring up naked or bare stalks loden with many hollow blew floures, of a strong sweet smell somewhat stuffing the head: after which come the cods or round knobs, containing a great quantitie of small blacke shining seed. The root is bulbous, full of a slimie glewish juice, which will serve to set feathers upon arrowes in stead of glew, or to paste bookes with: hereof is made the best starch next unto that of Wake-robin roots.

False bumbast Jacinth

The blew Hare-bels grow wilde in woods, Copses, and in the borders of fields every where thorow England. Our English Hyacinth is called *Hyacinthus Anglicus*, for that it is thought to grow more plentifully in England than elsewhere.

The roots, after the opinion of *Dioscorides*, being beaten and applied with white Wine, hinder or keepe backe the growth of haires.

¶ *Two Feigned Plants*

I have thought it convenient to conclude the historie of the Hyacinths with these two bulbous Plants, received by tradition from others, though generally holden for feigned and adulterine. Their pictures I could willingly have omitted in this historie, if the curious eye could elsewhere have found them drawne and described in our English Tongue: but because I finde them in none, I will lay them downe here, to the end that it may serve for excuse to others who shall come after, which list not to describe them, being as I said condemned for feined and adulterine nakedly drawne onely. The floures (saith the Author) are no lesse strange than wonderfull. The leaves and roots are like to those of Hyacinths. The floures resemble the Daffodils or Narcissus. The whole plant consisteth of a woolly or flockie matter: which description with the Picture was sent unto *Dodonæus* by *Johannes Aicholzius.*

The floure of Tygris

The second feigned picture hath beene taken of the Discoverer and others of late time, to be a kinde of Dragons not seene by any that have written thereof; which hath moved them to thinke it a feigned picture likewise; notwithstanding you shall receive the description thereof as it hath come to my hands. The root (saith my Author) is bulbous or Onion fashion, outwardly blacke; from the which spring up long leaves, sharpe pointed, narrow, and of a fresh greene colour: in the middest of which leaves rise up naked or bare stalkes, at the top whereof groweth a pleasant yellow floure, stained with many small red spots here and there confusedly cast abroad : and in the middest of the floure thrusteth forth a long red tongue or stile, which in time groweth to be the cod or seed-vessell, crooked or wreathed, wherein is the seed. The vertues and temperature are not to be spoken of, considering that we assuredly persuade our selves that there are no such plants, but meere fictions and devices, as we terme them, to give his friend a gudgeon.

WHITE LILLIES

The white Lilly hath long smooth and full bodied leaves of a grassie or light green colour. The stalks be two cubits high, and somtimes more, set or garnished with the like leaves, but growing smaller and smaller toward the top; and upon them do grow faire white floures strong of smell, narrow toward the foot of the stalke whereon they do grow, wide or open in the mouth like a bell. In the middle part of them doe grow small tender pointals tipped with a dusty yellow colour, ribbed or chamfered on the back side, consisting of six small leaves thicke and fat. The root is a bulb made of scaly cloves, full of tough and clammy juice, wherewith the whole plant doth generally abound.

The white Lilly of Constantinople hath very large & fat leaves like the former, but narrower and lesser. The stalke riseth up to the height of three cubits, set and garnished with leaves also like the precedent, but much lesse. Which stalke oftentimes doth alter and degenerat from his naturall roundnesse to a flat forme, as it were a lath of wood furrowed or chanelled alongst the same, as it were ribs or welts. The floures grow at the top like the former, saving that the leaves doe turne themselves more backward like the Turks cap, and beareth many more floures than our English white Lilly doth.

Our English white Lilly groweth in most gardens of England. The other groweth naturally in Constantinople and the parts adjacent, from whence we had plants for our English gardens, where they flourish as in their owne countrey.

The Lilly is called in Latine, *Rosa Junonis*, or *Juno's* Rose, because as it is reported it came up of her milke that fell upon the ground. But the Poets feign, That *Hercules*, who *Jupiter* had by *Alcumena*, was put to *Juno's* breasts whilest shee was asleepe; and after the sucking there fell away aboundance of milk, and that one part was spilt in the heavens, and the other upon the earth; and that of this sprang the Lilly, and of the other the circle in the heavens called *Lacteus Circulus*, or the Milky way, or otherwise in English Watling street. S. *Basil* in the explication of the 44 Psalm saith, That no floure so lively sets forth the frailty of mans life as the Lilly.

The root of the garden Lilly stamped with hony gleweth together sinues that be cut in sunder.

Florentinus a writer of Husbandry saith, That if the root be curiously opened, and therein be put some red, blew, or yellow colour that hath no causticke or burning qualitie, it will cause the floure to be of the same colour.

Mountaine Lillies

The great mountain Lilly hath a cloved bulb or scaly root, yellow of colour, very small in respect of the greatnesse of the plant; from the which riseth up a stalke, somtimes two or three, according to the age of the plant, whereof the middle stalke commonly turneth from his roundnesse into a flat forme, as those of the white Lilly of Constantinople. Upon these stalks do grow faire leaves of a blackish greene colour, in roundles and spaces as the leaves of Woodroofe, not unlike to the leaves of white Lilly, but smaller at the top of the stalkes. The floures be in number infinite, or at the least hard to be counted, very thicke set or thrust together, of an overworne purple, spotted on the inside with

The great mountaine Lilly

many smal specks of the colour of rusty iron. The whole floure doth turne it selfe backward at such time as the sun hath cast his beames upon it, like unto the Tulipa or Turks cap, as the Lilly or Martagon of Constantinople doth; from the middle whereof do come forth tender pendants hanging thereat, of the colour the floure is spotted with.

There hath not bin any thing left in writing either of the nature or vertues of these plants: notwithstanding we may deem, that God which gave them such seemely and beautifull shape, hath not left them without their peculiar vertues, the finding out whereof we leave to the learned and industrious searcher of Nature.

PERSIAN LILLY

The Persian Lilly hath for his root a great white bulbe, differing in shape from the other Lillies, having one great bulbe firme or solid, full of juice, which commonly each yere setteth off or encreaseth one other bulbe, and some-times more, which the next yere after is taken from the mother root, and so bringeth forth such floures as the old plant did. From this root riseth up a fat thicke and straight stem of two cubits high, whereupon is placed long narrow leaves of a greene colour, declining to blewnesse as doe those of the woad. The floures grow alongst the naked part of the stalk like little bels, of an overworn purple colour, hanging down their heads, every one having his own foot-stalke of two inches long, as also his pestell or clapper from the middle part of the floure; which being past and withered, there is not found any seed at all, as in other plants, but is encreased only in his root.

This Persian Lilly groweth naturally in Persia and those places adjacent, whereof it tooke his name, and is now (by the industry of Travellers into those countries, lovers of plants) made a denizon in some few of our London gardens.

There is not any thing known of the nature or vertues of this Persian Lilly, esteemed as yet for his rarenesse and comely proportion; although (if I might bee so bold with a stranger that hath vouchsafed to travell so many hundreds of miles for our acquaintance) wee have in our English fields many scores of floures in beauty far excelling it.

D AY   L I L L I E

The floures of the Day Lillie be like the white Lillie in shape, of an Orenge tawny colour: of which floures much might be said which I omit. But in briefe, this plant bringeth forth in the morning his bud, which at noone is full blowne, or spred abroad, and the same day in the evening it shuts it selfe, and in a short time after becomes as rotten and stinking as if it had beene trodden in a dunghill a moneth together, in foule and rainie weather: which is the cause that the seed seldome followes, as in the other of his kinde, not bringing forth any at all that I could ever observe; according to the old proverbe, Soone ripe, soone rotten.

These Lillies do grow in my garden, as also in the gardens of Herbarists, and lovers of fine and rare plants; but not wilde in England as in other countries. They do floure somewhat before the other Lillies.

It is fitly called, Faire or beautifull for a day: and so we in English may rightly tearme it the Day-Lillie, or Lillie for a day.

The roots and the leaves be laid with good successe upon burnings and scaldings.

T URKIE   OR   G INNY - HEN   F LOURE

The Checquered Daffodill, or Ginny-hen Floure, hath small narrow grassie leaves; among which there riseth up a stalke three hands high, having at the top one or two floures, and sometimes three, which consisteth of six small leaves checquered most strangely: wherein Nature, or rather the Creator of all things, hath kept a very wonderfull order, surpassing (as in all other things) the curiousest painting that Art can set downe. One square is of a greenish yellow colour, the other purple, keeping the same order as well on the backside of the floure as one the inside, although they are blackish in one square, and of a Violet colour in an other; insomuch that every leafe seemeth to be the feather of a Ginny hen, whereof it tooke his name. The root is small, white, and of the bignesse of halfe a garden beane.

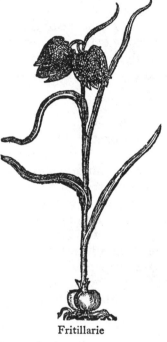

Fritillarie

The Ginny hen floure is called of *Dodonæus, Flos Meleagris*: of *Lobelius, Lilionarcissus variegata*, for that it hath the floure of a Lilly, and the root of *Narcissus*: it hath beene called *Fritillaria*, of the table or boord upon which men play at Chesse, which square checkers the floure doth very much resemble; some thinking that it was named *Fritillus*: whereof there is no certainty; for *Martial* seemeth to call *Fritillus, Abacus*, or the Tables whereon men play at Dice, in the fifth booke of his Epigrams, writing to *Galla.*

The sad Boy now his nuts cast by,
Is call'd to Schoole by Masters cry:

61

> And the drunke Dicer now betray'd
> By flattering Tables as he play'd,
> Is from his secret tipling house drawne out,
> Although the Officer he much besought, &c.

In English we may call it Turky-hen or Ginny-hen Floure, and also Checquered Daffodill, and Fritillarie, according to the Latine.

Of the facultie of these pleasant floures there is nothing set downe in the antient or later Writer, but they are greatly esteemed for the beautifying of our gardens, and the bosoms of the beautifull.

TULIPA, OR THE DALMATIAN CAP

Tulipa or the Dalmatian Cap is a strange and forrein floure, one of the number of the bulbed floures, whereof there be sundry sorts, some greater, some lesser, with which all studious and painefull Herbarists desire to be better acquainted, because of that excellent diversitie of most brave floures which it beareth. Of this there be two chiefe and generall kindes, *viz. Præcox*, and *Serotina*; the one doth beare his floures timely, the other later. To these two we will adde another sort called *Media*, flouring betweene both the others. And from these three sorts, as from their heads, all other kindes doe proceed, which are almost infinite in number. Notwithstanding, my loving friend Mr. *James Garret*, a curious searcher of Simples, and learned Apothecarie of London, hath undertaken to finde out, if it were possible, their infinite sorts, by diligent sowing of their seeds, and by planting those of his owne propagation, and by others received from his friends beyond the seas for the space of twenty yeares, not being yet able to attaine to the end of his travel, for that each new yeare bringeth forth new plants of sundry colours not before seen; all which to describe particularly were to rolle *Sisiphus* stone, or number the sands. So that it shall suffice to speake of and describe one, referring the rest to some that meane to write of Tulipa a particular volume.

The Tulipa of Bolonia hath fat thicke and grosse leaves, hollow, furrowed or chanelled, bended a little

backward, and as it were folded together: which at their first comming up seeme to be of a reddish colour, and being throughly growne turne into a whitish greene. In the midst of those leaves riseth up a naked fat stalke a foot high, or somthing more; on the top whereof standeth one or two yellow floures, somtimes three or more, consisting of six smal leaves, after a sort like to a deepe wide open cup, narrow above, and wide in the bottome. After it hath been some few dayes floured, the points and brims of the floure turn backward, like a Dalmatian or Turkish Cap, called Tulipan, Tolepan, Turban, and Turfan, whereof it tooke his name. The chives or threds in the middle of the floure be somtimes yellow, otherwhiles blackish or purplish, but commonly of one overworne colour or other, Nature seeming to play more with this floure than with any other that I do know. This floure is of a reasonable pleasant smell, and the other of his kinde have little or no smel at all. The root is bulbous, and very like to a common onion of S. *Omers*.

Clusius his greater Tulip

We have likewise another of greater beautie, and very much desired of all, with white floures dasht on the backside, with a light wash of watched colour.

Tulipa groweth wilde in Thracia, Cappadocia, and Italy; in Bizantia about Constantinople; at Tripolis and Aleppo in Syria. They are now common in all the English gardens of such as affect floures.

They floure from the end of Februarie unto the beginning of May, and somwhat after: although *Augerius Busbequius* in his journey to Constantinople, saw betweene Hadrianople and Constantinople, great aboundance of them in floure everie where, even in the midst of Winter, in the moneth of Januarie, which that warme and temperat clymat may seeme to performe.

The later Herbarists by a Turkish or strange name call it Tulipa, of the Dalmatian cap called Tulipa, the forme whereof the floure when it is open seemeth to represent.

It is called in English after the Turkish name Tulipa, or it may be called Dalmatian Cap, or the Turks Cap. What name the antient Writers gave it is not certainly knowne.

Floure de-luce

The common Floure de-luce hath long and large flaggy leaves like the blade of a sword with two edges, amongst which spring up smooth and plaine stalks two foot long, bearing floures toward the top compact of six leaves joyned together, whereof three that stand upright are bent inward one toward another; and in those leaves that hang downeward there are certaine rough or hairy welts, growing or rising from the nether part of the leafe upward, almost of a yellow colour. The roots be thicke, long, and knobby, with many hairy threds hanging thereat.

The water Floure de-luce, or water Flag, or *Acorus*, is like unto the garden Floure de-luce in roots, leaves, and stalkes, but the leaves are much longer, sometimes of the height of foure cubits, and altogether narrower. The floure is of a perfect yellow colour, and the root knobby like the other; but being cut, it seemes to be of the colour of raw flesh. The water Floure de-luce or yellow Flag prospereth well in moist medowes, and in the borders and brinks of Rivers, ponds, and standing lakes. Although it be a watery plant of nature, yet being planted in gardens it prospereth well.

Turky Floure de-luce hath long and narrow leaves of a blackish green like stinking Gladdon; among which rise up stalks two foot long, bearing at the top of each stalke one floure compact of six great leaves: the three that stand upright are confusedly and very strangely striped, mixed with white and a duskish blacke colour. The three leaves that hang downward are like a gaping hood, and are mixed in like manner, (but the white is nothing so bright as of the other) and are as it were shadowed over with a darke purple colour somwhat shining; so that according to my judgment, the whole floure is of the colour of a Ginny hen, a rare and beautifull floure to behold.

Turky Floure de-luce

The Floure de-luce of Florence, whose root in shops and generally every where are called *Ireos*, or *Orice* (whereof sweet waters, sweet pouders, and such like are made) is altogether like unto the common Floure de-luce, saving that the floures of the *Ireos* is of a white colour, and the roots exceeding sweet of smell, and the other of no smell at all.

The great Floure de-luce of Dalmatia hath leaves much broader, thicker, and more closely compact together than any of the other, and set in order like wings or the fins of a Whale fish, greene toward the top, and of a shining purple colour toward the bottome, even to the ground: amongst which riseth up a stalke of foure foot high, as my selfe did measure oft times in my garden: whereupon doth grow faire large floures of a light blew, or as we terme it a watchet colour. The floures do smell exceeding sweet, much like the Orenge floure. The root hath no smell at all.

The root of the common Floure de-luce cleane washed, and stamped with a few drops of Rose-water, and laid plaisterwise upon the face of man or woman, doth in two daies at the most take away the blacknesse or blewnesse of any stroke or bruse; so that if the skinne of the same woman or any other person be very tender and delicate, it shall be needfull that ye lay a piece of silke, sindall, or a piece of fine laune betweene the plaister and the skinne; for otherwise in such tender bodies it often causeth heat and inflammation. There is an excellent oyle made of floures and roots of Floure de-luce, of each a like quantitie, called *Oleum Irinum*, made after the same manner that oyle of Roses, Lillies and such like be made: which oyle profiteth much to strengthen the sinewes and joynts, helpeth the crampe proceeding of repletion, and the disease called in Greeke *Peripneumonia*.

Peionie

Peionie hath thicke red stalkes a cubit long: the leaves be great and large, consisting of divers leaves growing or joyned together upon one slender stemme or rib, not

The double Peionie

much unlike the leaves of the Wallnut tree both in fashion and greatnesse: at the top of the stalkes grow faire large redde floures very like roses, having also in the midst, yellow threds or thrums like them in the rose called *Anthera*; which being vaded and fallen away there come in place three or foure great cods or husks, which do open when they are ripe; wherein is contained blacke shining and polished seeds, as big as a Pease.

Apuleius saith, that the seeds or graines of Peionie shine in the night time like a candle, and that plenty of it is in the night season found out and gathered by the shepheards.

Ælianus saith, that it is not plucked up without danger; and that it is reported how he that first touched it, not knowing the nature thereof, perished. Therefore a string must be fastned to it in the night, and a hungry dog tied therto, who being allured by the smell of rosted flesh set towards him, may plucke it up by the roots.

Moreover, it is set downe by the said Author, that of necessitie it must be gathered in the night; for if any man shall pluck off the fruit in the day time, being seene of the Wood-pecker, he is in danger to lose his eies. The like fabulous tale hath been set forth of Mandrake. But all these things be most vaine and frivolous: for the root of Peionie, as also the Mandrake, may be removed at any time of the yeare, day or houre whatsoever.

But it is no marvell, that such kindes of trifles, and most superstitious and wicked ceremonies are found in the books of the most Antient Writers; for there were many things in their time very vainly feined and cogged in for ostentation sake, as by the Ægyptians and other counterfeit mates, as *Pliny* doth truly testifie. It is reported that these herbes tooke the name of Peionie, or *Pæan*, of that excellent Physition of the same name, who first found out and taught the knowledge of this herbe unto posteritie.

CORNE-FLAG

French Corne-flagge hath small stiffe leaves ribbed or chamfered with long nerves or sinues running through the same, in shape like those of the small Floure de-luce, or the blade of a sword, sharpe pointed, of an overworne green colour, among which riseth up a stif brittle stalk two cubits high, wherupon do grow in comly order many faire purple flours gaping like those of Snapdragon, or not much differing from the Fox-glove called in Latine *Digitalis*. After them come round knobby seed-vessesl full of chaffie seed, very light, of a brown reddish colour. The root consists of two bulbes one set upon the other; the uppermost whereof in the beginning of the spring is lesser, and more full of juice; the lower greater, but more loose and lithie, which shortly after perisheth.

These kindes of Corne-flags grow in medowes and in earable grounds among corne, in many places of Italy, as also in the parts of France bordering thereunto. Neither are the fields of Austria and Moravia without them. We have great plenty of them in our London gardens, especially for the garnishing and decking them up with their seemly flowers.

The floures of the Corne-flag are called of the Italians, *Monacuccio*: in English, Corne-Flag, Corne-Sedge, Sword-Flag, Corne Gladin: in French, *Glais*.

The root stamped with the pouder of Frankincense and wine, and applied, draweth forth splinters and thornes that sticke fast in the flesh.

The cods with the seed dried and beaten into pouder, and drunk in Goats milke or Asses milke, presently taketh away the paine of the Colique.

COLUMBINE

The blew Columbine hath leaves like the great Celandine, but somewhat rounder, indented on the edges, parted into divers sections, of a blewish green colour, which beeing broken, yeeld forth little juice or none at all: the stalke is a cubit and a halfe high, slender, reddish, and sleightly haired: the slender sprigs whereof bring forth everie one one floure with five little hollow hornes, as it were hanging forth, with small leaves standing upright, of the shape of little birds: these floures are of colour somtimes blew, at other times of a red or purple, often white, or of mixt colours, which to distinguish severally were to small purpose, being things so familiarly known to all: after the floures grow up cods, in which is contained little black and glittering seed: the roots are thicke, with some strings thereto belonging, which continue many yeres.

Columbine

69

They are set and sowne in gardens for the beautie and variable colour of the floures.

Columbine is called of the later Herbarists, *Aquilegia*: of some, *Herba Leonis*, or the herbe wherein the Lion doth delight. They are used especially to decke the gardens of the curious, garlands and houses.

Calves snout, or Snapdragon

The purple Snapdragon hath great and brittle stalks, which divideth it selfe into many fragile branches, whereupon do grow long leaves sharpe pointed, very greene, like unto those of wilde flax, but much greater, set by couples one opposite against another. The floures grow at the top of the stalkes, of a purple colour, fashioned like

a frogs mouth, or rather a dragons mouth, from whence the women have taken the name Snapdragon. The seed is blacke, contained in round huskes fashioned like a calves snout, (whereupon some have called it Calves snout) or in mine opinion it is more like unto the bones of a sheeps head that hath beene long in the water, or the flesh consumed cleane away.

The second agreeth with the precedent in every part, except in the colour of the floures, for this plant bringeth forth white floures, and the other purple, wherein consists the difference.

Snapdragon

The yellow Snapdragon hath a long thicke wooddy root, with certain strings fastned thereto ; from which riseth up a brittle stalke of two cubits and a halfe high, divided from the bottome to the top into divers branches, whereupon do grow long greene leaves like those of the former, but greater and longer. The floures grow at the top of the maine branches, of a pleasant yellow colour, in shape like unto the precedent.

That which hath continued the whole Winter doth floure in May, and the rest of Summer afterwards; and that which is planted later, and in the end of Summer, floureth in the Spring of the following yeare: they do hardly endure the injurie of our cold Winter.

Snapdragon is called in English, Calves snout, Snapdragon, and Lyons snap: in French, *Teste de chien*, and *Teste de Veau*.

They report (saith *Dioscorides*) that the herbe being hanged about one preserveth a man from being bewitched, and that it maketh a may gracious in the sight of people.

Peach-bells and Steeple-bells

The Peach-leaved Bell-floure hath a great number of small and long leaves, rising in a great bush out of the ground, like the leaves of the Peach-tree: among which riseth up a stalke two cubits high: alongst the stalke grow many floures like bells, sometime white, and for the most part of a faire blew colour; but the bells are

Peach-bells

nothing so deepe as they of the other kindes; and these are more dilated or spread abroad than any of the rest. The seed is small like Rampions, and the root a tuft of laces or small strings.

The second kinde of Bell-floure hath a great number of faire Blewish or Watchet floures, like the other last before mentioned, growing upon goodly tall stems two cubits and a halfe high, which are garnished from the top of the plant unto the ground with leaves like Beets, disorderly placed. This whole plant is exceeding full of milke, insomuch as if you do but breake one leafe of the plant, many drops of a milky juice will fall upon the ground. The root is very great, and full of milke also: likewise the knops wherein the seed should be are empty and void of seed, so that the whole plant is altogether barren, and must be increased with slipping of his root.

These Bell-floures grow in our London Gardens, and not wilde in England.

CRANES-BILL

Doves-foot hath many hairy stalks, trailing or leaning toward the ground, of a brownish colour, somewhat kneed or joynted; wherupon do grow rough leaves of an overworn green color, round, cut about the edges, and like unto those of the common Mallow: amongst which come forth the floures of a bright purple colour: after which is the seed, set together like the head and bill of a bird; wherupon it was called Cranes-bill, or Storks-bill.

It is found neere to common high waies, desart places, untilled grounds, and specially upon mud walls almost every where.

It is commonly called in Latine, *Pes Columbinus*: in French, *Pied de Pigeon*: hereupon it may be called *Geranium Columbinum*: in English, Doves-foot, and Pigeons foot.

Doves-foot

The herbe and roots dried, beaten into most fine pouder, and given halfe a spoonfull fasting, and the like quantitie to bedwards in red wine, or old claret, for the space of one and twenty daies together, cure miraculously ruptures or burstings, as my selfe have often proved, whereby I have gotten crownes and credit: if the ruptures be in aged persons, it shall be needfull to adde thereto the powder of red snailes (those without shels) dried in an oven in number nine, which fortifieth the herbes in such sort, that it never faileth, although the rupture be great and of long continuance: it likewise profiteth much those that are wounded into the body, and the decoction of the herbe made in wine, prevaileth mightily in healing inward wounds, as my selfe have likewise proved.

Thrift, or our Ladies Cushion

Thrift is a kinde of Gillofloure, which brings forth leaves in great tufts, thicke thrust together, among which rise up small tender stalkes of a spanne high, naked and without leaves; on the tops whereof stand little floures in a spokie tuft, of a white colour tending to purple.

Thrift is found in the most salt marshes in England, as also in Gardens, for the bordering up of beds and bankes, for the which it serveth very fitly.

73

They floure from May, till Summer be farre spent. Thrift is called in English, Thrift, Sea-grasse, and our Ladies Cushion.

Their use in Physicke as yet is not knowne, neither doth any seeke into the Nature thereof, but esteeme them only for their beautie and pleasure.

CROW-FEET

There be divers sorts or kinds of these pernitious herbes comprehended under the name of *Ranunculus*, or Crow-foot, whereof most are very dangerous to be taken into the body, and therefore they require a very exquisite moderation, with a most exact and due maner of tempering, not any of them are to be taken alone by themselves, because they are of most violent force, and

therefore have the greater need of correction. These dangerous simples are likewise many times of themselves beneficial, & oftentimes profitable: for some of them are not so daungerous, but that they may in some sort, and oftentimes in fit and due season profit and doe good.

The common Crow-foot hath leaves divided into many parts, commonly three, sometimes five, cut here and there in the edges, of a deep green colour, in which stand divers white spots: the stalks be round, somthing hairie, some of them bow downe toward the ground, and put forth many little roots, whereby it taketh hold of the

Common Crow-foot

74

ground as it traileth along: some of them stand upright, a foot high or higher; on the tops whereof grow small flours with five leaves apiece, of a yellow glittering colour like gold. Crow-foot is called in English King Kob, Gold cups, Gold knobs, Crow-foot, and Butter-floures.

Many do use to tie a little of the herbe stamped with salt unto any of the fingers, against the pain of the teeth; which medicine seldome faileth; for it causeth greater paine in the finger than was in the tooth, by the meanes whereof the greater paine taketh away the lesser.

Cunning beggers do use to stampe the leaves, and lay it unto their legs and arms, which causeth such filthy ulcers as we dayly see (among such wicked vagabonds) to move the people the more to pittie.

Crow-foot of Illyria spoileth the sences and understanding, and draweth together the sinewes and muscles of the face in such strange manner, that those who beholding such as died by the taking hereof, have supposed that they died laughing; so forceably hath it drawne and contracted the nerves and sinewes, that their faces have beene drawne awry, as though they laughed, whereas contrariwise they have died with great torment.

MEDOW TREFOILE

Medow Trefoile bringeth forth stalkes a cubit long, round and something hairy, the greater part of which creepeth upon the ground: whereon grow leaves consisting of three joined together, one standing a little from another. The floures grow at the tops of the stalks in a tuft or small Fox-taile eare, of a purple colour, and sweet of taste.

Common medow Trefoile grows in medowes, fertile pastures, and waterish grounds.

Meadow Trefoile is called in English Common Trefoile, Three leafed grasse: of some, Suckles, Hony-suckles, and Cocks-heads: in Irish, *Shamrocks.*

75

Medow Trefoile

Oxen and other cattell do feed on the herb, as also calves and yong lambs. The flours are acceptable to Bees.

Pliny writeth and setteth it downe for certaine, that the leaves hereof do tremble and stand right up against the comming of a storme or tempest.

HERBE TWO-PENCE

Herbe Two-pence hath a smal and tender root, spreding and dispersing it selfe far within the ground, from which rise up many little, tender, flexible stalks trailing upon the ground, set by couples at certaine spaces, with smooth greene leaves somewhat round, whereof it tooke his name: from the bosome of which leaves shoot forth small tender foot-stalks, whereon do grow little yellow floures, like those of Cinkefoile or Tormentill.

It groweth neere unto ditches and streames, and other waterie places, and is somtimes found in moist woods: I found it upon the banke of the river of Thames, right against the Queenes palace of White-hall; and almost in every countrey where I have travelled.

It floureth from May till Summer be well spent.

Herb Two-pence is called in Latine *Nummularia* and *Centummorbia*: and of divers *Serpentaria*. It is reported, that if serpents be hurt or wounded, they do heale themselves with this herb, wherupon came the name *Serpentaria*: and it is called *Nummularia* of the forme of money, whereunto the leaves are like: in English, Money-woort, Herbe Two-pence, and Two-penny grasse.

The floures and leaves stamped and laid upon wounds and ulcers do cure them: but it worketh most effectually being stamped and boiled in oile olive, with some rosin, wax, and turpentine added thereto.

Boiled with wine and hony it cureth the wounds of the inward parts, and ulcers of the lungs; & in a word, there is not a better wound herb, no not Tabaco it selfe, nor any other whatsoever.

The herb boiled in wine, with a little hony or mead, prevaileth much against the cough in children, called the Chin-cough.

Herbe Two-pence

NEESING ROOT OR NEESEWORT

White Hellebor hath leaves like unto great Gentian, but much broader, and not unlike the leaves of the great Plantaine, folded into pleits like a garment plaited to be laid up in a chest: amongst these leaves riseth up a stalke a cubit long, set towards the top full of little star-like floures of an herby green colour tending to whitenes: which being past, there come small husks containing the seed. The root is great and thicke, with many small threds hanging thereat.

The white Hellebor groweth on the Alps and such like

77

mountains where Gentian groweth. It was reported unto me by the Bishop of Norwich, That white Hellebor groweth in a wood of his owne neere to his house at Norwich. Some say likewise that it doth grow upon the mountaines of Wales. I speake this upon report, yet I thinke it may be true. Howbeit I dare assure you that they grow in my garden at London.

The root of white Hellebor is good against phrensies, sciatica, dropsies, poison, and against all cold diseases that be of hard curation. This strong medicine made of white Hellebor, ought not to bee given inwardly unto delicate bodies without great correction; but it may be more safely given unto countrey people which feed grosly, and have hard tough and strong bodies.

The pouder drawne up into the nose causeth sneesing, and purgeth the brain from grosse and slimie humors.

The root killeth mice and rats, being made up with hony and floure of wheat.

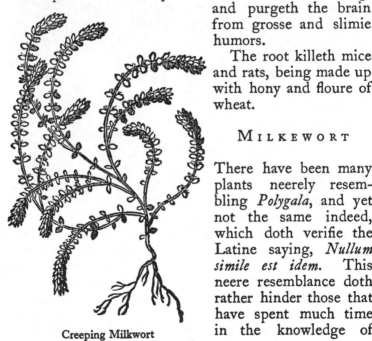

Milkewort

There have been many plants neerely resembling *Polygala*, and yet not the same indeed, which doth verifie the Latine saying, *Nullum simile est idem*. This neere resemblance doth rather hinder those that have spent much time in the knowledge of

Creeping Milkwort

Simples, than increase their knowledge: and this also hath been an occasion that many have imagined a sundrie *Polygala* unto themselves, and so of other plants. Of which number this whereof I speake is one, obtaining this name of the best writers and herbarists of our time, describing it thus: It hath many thicke spreading branches creeping on the ground, bearing leaves like those of *Herniaria*, standing in rowes like the sea Milkwort; among which grow smal whorles or crownets of white floures, the root being exceeding small and threddy.

The second kinde of *Polygala* is a small herbe with pliant slender stemmes, of a wooddy substance, an handfull long, creeping by the ground : the leaves be small and narrow like to Lintels, or little Hyssop. The floures grow at the top, of a blew colour, fashioned like a little bird, with wings, taile, and body easie to be discerned by them that do observe the same: which being past, there succeed small pouches like those of *Bursa pastoris*, but lesser. The root is small and wooddy.

This third kinde of *Polygala* or Milkwort, hath leaves and stalkes like the last before mentioned, and differeth from it only herein, that this kinde hath smaller branches, and the leaves are not so thicke thrust together, and the floures are like the other, but that they be of a red or purple colour.

The fourth kinde is like the last spoken of in every respect, but that it hath white floures, otherwise it is very like.

Purple Milkewort differeth from the others in the colour of the floures, it bringeth forth moe branches than the precedent, and the floures are of a purple colour, wherin especially consists the difference.

The sixt Milkwort is like unto the rest in each respect, saving that the floures are of an overworne ilfavored colour, which maketh it to differ from all the other of his kinde.

These plants or Milke-worts grow commonly in every wood or fertil pasture wheresoever I have travelled.

Milkwort is called by *Dodonæus*, *Flos Ambarualis*, because it doth especially floure in the Crosse or Gang weeke, or Rogation weeke: of which floures the maidens which use in the countries to walke the Procession do make themselves garlands and Nosegaies: in English wee may call it Crosse-floure, or Procession floure, Gang-floure, Rogation-floure, and Milkwort, of their vertues in procuring milke in the brests of nurses.

ARCHANGELL OR DEAD NETTLE

White Archangell hath foure square stalkes a cubit high, leaning this way and that way, by reason of the great weight of his ponderous leaves, which are in shape like unto those of Nettles, nicked round about the edges, yet not stinging at all, but soft and as it were downy: the floures compasse the stalkes round about at certaine distances.

Yellow Archangell hath square stalks rising from a threddy root, set with leaves by couples very much cut or hackt about the edges, and sharp pointed, the uppermost whereof are oftentimes of a faire purple colour: the flours grow among the said leaves, of a gold yellow colour, fashioned like those of the white Archangell, but greater, and wider gaping open.

White Archangell

Red Archangell, being called *Urtica non mordax*, or dead Nettle, hath many leaves spred upon the ground; among which rise up stalkes hollow and square, whereupon grow rough leaves of an overworne colour, among which come forth purple floures set about in round wharles or rundles. The root is small, and perisheth at the first approch of winter.

These plants are found under hedges, old walls, common waies, among rubbish, in the borders of fields, and in earable grounds, oftentimes in gardens ill husbanded.

That with the yellow floure groweth not so common as the others. I have found it under the hedge on the left hand as you go from the village of Hampsted neer London to the Church, & in the wood therby, as also in many other copses about Lee in Essex, neer Watford & Bushy in Middlesex, and in the woods belonging to the Lord Cobham in Kent.

They floure for the most part all Summer long, but chiefely in the beginning of May.

Archangell is called in English, Archangell, blinde Nettle, and dead Nettle.

The floures are baked with sugar as Roses are, which is called Sugar roset: as also the distilled water of them, which is used to make the heart merry, to make a good colour in the face, and to refresh the vitall spirits.

BLACK BRIONIE, OR THE WILDE VINE

The black Briony hath long flexible branches of a woody substance, covered with a gaping or cloven bark growing very far abroad, winding it selfe with his small tendrels about trees, hedges, and what else is next unto it, like unto the branches of the vine: the leaves are like unto those of Ivie or garden Nightshade, sharpe pointed, and of a shining greene colour: the floures are white, small, and mossie; which being past, there succeed little clusters of red berries, somewhat bigger than those of the small Raisins or Ribes, which wee call Currans or small Raisins. The root is very great and thicke, oftentimes as big as a mans leg, blackish without, and very clammy or slimy within; which beeing but scraped with a knife or any other thing fit for that purpose, it seemes to be a matter fit to spred upon cloath or leather in manner of a plaister of Sear-cloath: which being so spred and used, it serveth to lay upon many infirmities, and unto verie excellent purposes, as shal be declared.

These plants grow in hedges and bushes almost every where. They spring in March, bring forth the floures in May, and their ripe fruit in September.

The yong and tender sproutings are kept in pickle, and reserved to be eaten with meat, as *Dioscorides* teacheth. *Matthiolus* writeth, that they are served at mens tables in our age also in Tuscanie; others also report the like to be done in Andolosia one of the kingdomes of Granado.

The roots spred upon sheeps leather in manner of a plaister, whilest it is yet fresh and greene, taketh away blacke and blew marks, all scars and deformitie of the skin, breaks hard apostems, drawes forth splinters and broken bones, dissolveth congealed bloud, and being laid on and used upon the hip or huckle bones, shoulders, armes, or any other part where there is great paine and ache, it takes it away in short space, and worketh very effectually.

BINDE-WEED

The common rough Bind-weed hath many branches set full of little sharpe prickles, with certaine clasping tendrels, wherewith it taketh hold upon hedges, shrubs, and whatsoever standeth next unto it, winding and clasping it selfe about from the bottome to the top; whereon are placed at every joint one leafe like that of Ivie, with-

Common rough Bindeweed

out corners, sharpe pointed, lesser and harder than those of smooth Binde-weed, oftentimes marked with little white spots, and garded or bordered about the edges with crooked prickles. The floures grow at the top of crooked stalks of a white colour, and sweet of smell. After commeth the fruit like those of the wilde Vine, greene at the first, and red when they be ripe, and of a biting taste; wherein is contained a blackish seed in shape like that of hempe. The root is long, somewhat hard, and parted into very many branches.

It is a strange thing unto me, that the name of *Smilax* should be so largely extended, as that it should be assigned to those plants that come nothing neere the nature, and scarsly unto any part of the forme of *Smilax* indeed. But we will leave controversies to the further consideration of such as love to dance in quag-mires, and come to this our common smooth *Smilax*, called and knowne by that name among us, or rather more truly by the name of *Convolvulus major*, or *Volubilis major*: it beareth the long branches of a Vine, but tenderer, and for the length and great spreading thereof it is very fit to make shadows in arbors: the leaves are smooth like Ivie, but somwhat bigger, and being broken are full of milke: amongst which come forth great white and hollow floures like bells.

Scammonie of Syria or purging Bindweed hath many

Syrian Scammonie

stalkes rising from one root, which are long, slender, and like the clasping tendrels of the vine, by which it climeth and taketh hold of such things as are next unto it. The leaves be broad, sharpe pointed like those of the smooth or hedge Bind-weed: among which come forth very faire white floures tending to a blush colour, bell-fashion. The root is long, thicke, and white within: out of which is gathered a juyce that being hardned, is greatly used in Physicke: for which consideration, there is not any plant growing upon the earth, the knowledge whereof more concerneth a Physition, both for his shape and properties, than this Scammonie, which *Pena* calleth *Lactaria scansoriaque volvula*, that is, milky and climbing Windweed, whereof it is a kinde. And although this herbe be suspected, and halfe condemned of some learned men, yet there is not any other herbe to be found, whereof so small a quantity will do so much good: neither could those which have carped at it, and reproved this herbe, finde any simple in respect of his vertues to be put in his roome: and hereof ensueth great blame to all practitioners, who have not endevoured to be better acquainted with this herbe, chiefely to avoid the deceit of the crafty Drug-seller and Medicine-maker of this confected Scammony, brought us from farre places, rather to be called I feare infected Scammony, or poysoned Scammony, than confected.

84

Although we have great plenty of the roots of Binde-weed of Peru, which we usually cally *Zarza*, or *Sarsaparilla*, wherewith divers griefes and maladies are cured, and that these roots are very well knowne to all; yet such hath beene the carelessnesse and small providence of such as have travelled into the Indies, that hitherto not any have given us instruction sufficient, either concerning the leaves, floures, or fruit: onely *Monardus* saith, that it hath long roots deepe thrust into the ground: which is as much as if a great learned man should tell the simple, that our common carrion Crow were of a blacke colour. For who is so blinde that seeth the root it selfe, but can easily affirme the root to be very long? Notwithstanding, there is in the reports of such as say they have seene the plant it selfe growing, some contradiction or contrarietie: some report that it is a kind of Bindweed, and especially one of the rough Bindweeds.

Zarzaparilla of Peru is a strange plant, and is brought unto us from the Countries of the new world called America; and such things as are brought from thence, although they also seeme and are like to those that grow in Europe, notwithstanding they do often differ in vertue and operation: for the diversitie of the soile and of the weather doth not only breed an alteration in the forme but doth most of all prevaile in making the vertues and qualities greater or lesser. Such things as grow in hot places be of more force, and greater smell; and in cold, of lesser. Some things that are deadly and pernitious, being removed wax milde, and are made wholesome: so in like manner, although *Zarzaparilla* of Peru be like to rough Bind-weed, or to Spanish *Zarza parilla*, notwithstanding by reason of the temperature of the weather, and also through the nature of the soile, it is of a great deale more force than that which groweth either in Spaine or in Africke.

The roots are a remedie against long continuall paine

of the joynts and head, and against cold diseases. They are good for all manner of infirmities wherein there is hope of cure by sweating, so that there be no ague joyned.

WOOD-BINDE, OR HONY-SUCKLE

Wood-binde or Hony-suckle climeth up aloft, having long slender wooddy stalkes, parted into divers branches: about which stand by certaine distances smooth leaves, set together by couples one right against another; of a light greene colour above, underneath of a whitish greene. The floures shew themselves in the tops of the branches, many in number, long, white, sweet of smell, hollow within; in one part standing more out, with certaine threddes growing out of the middle. The fruit is like little bunches of grapes, red when they be ripe, wherein is contained small hard seed.

The Woodbinde groweth in woods and hedges, and upon shrubs and bushes, oftentimes winding it selfe so straight and hard about, that it leaveth his print upon those things so wrapped.

The double Honisuckle groweth now in my Garden, and many others likewise in great plenty, although not long since, very rare and hard to be found, except in the garden of some diligent Herbarists.

The leaves come forth be-times in the spring: the

Woodbinde or Honisuckle

floures bud forth in May and June: the fruit is ripe in Autumne.

The floures steeped in oile, and set in the Sun, are good to annoint the body that is benummed, and growne very cold.

SWALLOW-WORT

Swallow-wort with white floures hath divers upright branches of a brownish colour, of the height of two cubits, beset with leaves not unlike to those of *Dulcamara* or wooddy Night-shade, somewhat long, broad, sharpe pointed, of a blackish greene colour, and strong savor; among which come forth very many small white floures star-fashion, hanging upon little slender foot-stalkes: after which come in place thereof long sharp pointed cods, stuffed full of a most perfect white cotton resembling silk, as well in shew as handling (our London Gentlewomen have named it Silken Cislie), among which is wrapped soft brownish seed. The roots are very many, white, threddie, and of a strong savour.

It is called of the later Herbarists *Vincetoxicum*: in English, Swallow-woort. *Æsculapius* (who is said to be the first inventor of physick, whom therefore the Greekes and Gentiles honored as a god) called it after his owne name *Asclepias* or *Æsculapius* herbe, for that he was the first that wrote thereof, and now it is called in shops *Hirundinaria*.

Dioscorides writeth, That the roots of *Asclepias* or Swallow-woort boiled in wine, and the decoction drunke, are a remedy against the stingings of Serpents, and against deadly poyson, being one of the especiallest herbes against the same.

There groweth in that part of Virginia, or Norembega, where our English men dwelled (intending there to erect a certaine Colonie) a kinde of *Asclepias*, or Swallow-woort, which the Savages call *Wisanck*: there riseth up from a single crooked root, one upright stalk a foot high, slender, and of a greenish colour: whereupon do grow faire broad leaves sharp pointed, with many ribs or nerves running through the same like those of Ribwort or Plaintaine, set together by couples at certaine distances. The floures come forth at the top of the stalks, which as yet are not observed by reason the man that brought the seeds & plants hereof did not regard them: after which, there come in place two cods (seldome more) sharp pointed like those of our Swallow-woort, but greater, stuffed full of a most pure silke of a shining white colour: among which silke appeareth a small long tongue (which is the seed) resembling the tongue of a bird, or that of the herbe called Adders tongue. The cods are not only full of silke, but every nerve or sinew wherewith the leaves be ribbed are likewise most pure silke; and also the pilling of the stems, even as flax is torne from his stalks. This considered, behold the justice of God, that as he hath shut up those people and nations in infidelity and nakednes, so hath he not as yet given them understanding to cover their nakednesse, nor mater wherewith to do the same; notwithstanding the earth is covered over with this silke, which daily they tread under their feet, which were sufficient to apparell many kingdomes, if they were carefully manured and cherished.

It groweth, as before is rehearsed, in the countries of Norembega, now called Virginia, by the honourable Knight Sir *Walter Raleigh*, who hath bestowed great sums of money in the discoverie thereof; where are dwelling at this present English men.

SOLOMONS SEALE

Solomons Seale hath long round stalks, set for the most part with long leaves somewhat furrowed and ribbed, not much unlike Plantain, but narrower, which for the most part stand all upon one side of the stalke, and hath small white floures resembling the floures of Lilly Conval: on the other side when the floures be vaded, there come forth round berries, which at the first are green and of a blacke colour tending to blewnesse, and being ripe, are of the bignesse of Ivy berries, of a very sweet and pleasant taste. The root is white and thicke, full of knobs or joints, in some places resembling the mark of a seale, whereof I thinke it tooke the name *Sigillum Solomonis*; it is sweet at the first, but afterward of a bitter taste with some sharpnesse.

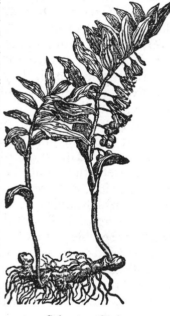

Dioscorides writeth, That the roots are excellent good for to seale or close up greene wounds, being stamped and laid thereon; whereupon it was called *Sigillum Salomonis*, of the singular vertue that it hath in sealing or healing up wounds, broken bones, and such like. Some have thought it tooke the name *Sigillum* of the markes upon the roots:

Solomons Seale

but the first reason seemes to be more probable.

The root of Solomons seale stamped while it is fresh and greene, and applied, taketh away in one night, or two at the most, any bruise, blacke or blew spots gotten by fals or womens wilfulnesse, in stumbling upon their hasty husbands fists, or such like.

Galen saith, that neither herbe nor root hereof is to be given inwardly: but note what experience hath found out, and of late daies, especially among the vulgar sort of people in Hampshire, which *Galen, Dioscorides,* or any other that have written of plants have not so much as dreamed of; which is, That if any of what sex or age soever chance to have any bones broken, in what part of their bodies soever; their refuge is to stampe the roots hereof, and give it unto the patient in ale to drinke: which sodoreth and glues together the bones in very short space, and very strangely, yea although the bones be but slenderly and unhandsomely placed and wrapped up. Moreover, the said people do give it in like manner unto their cattell, if they chance to have any bones broken, with good successe; which they do also stampe and apply outwardly in manner of a pultesse, as well unto themselves as their cattell.

The root stamped and applied in manner of a pultesse, and laid upon members that have beene out of joynt, and newly restored to their places, driveth away the paine, and knitteth the joynt very firmely, and taketh away the inflammation, if there chance to be any.

That which might be written of this herbe as touching the knitting of bones, and that truely, would seeme unto some incredible; but common experience teacheth, that in the world there is not to be found another herbe comparable to it for the purposes aforesaid: and therefore in briefe, if it be for bruises inward, the roots must be stamped, some ale or wine put thereto, strained, and given to drinke.

LINE OR LINDEN TREE

The female Line or Linden tree waxeth very great and thicke, spreading forth his branches wide and farre abroad, being a tree which yeeldeth a most pleasant shadow, under and within whose boughes may be made brave summer houses and banqueting arbors, because the more that it is surcharged with weight of timber and such like, the better it doth flourish. The barke is brownish, very smooth, and plaine on the outside, but that which is next to the timber is white, moist and tough, serving very well for ropes, trases, and halters. The timber is whitish, plaine and without knots, yea very soft and gentle in the cutting or handling. Better gunpouder is made of the coales of this wood than of Willow coales. The leaves are greene, smooth, shining, and large, somewhat snipt or toothed about the edges: the floures are littlè, whitish, of a good savour, and very many in number, growing clustering together from out of the middle of the leafe: out of which proceedeth a small whitish long narrow leafe: after the floures succeed cornered sharpe pointed Nuts, of the bignesse of Hasell Nuts.

The male *Tilia* or Line tree groweth also very great and thicke, spreading it selfe far abroad like the other Linden tree: his barke is very tough and pliant, and serveth to make cords and halters of. The timber of this tree is much harder, more knotty, and more yellow than the timber of the other, not much differing from the timber of the Elme tree: the leaves hereof are not much unlike Ivy leaves, not very greene, somewhat snipt about the edges: from the middle whereof come forth clusters of little white flours like the former: which being vaded, there succeed small round pellets, growing clustering together like Ivy berries, within which is contained a little round blackish seed, which falleth out when the berry is ripe.

The female Linden tree groweth in some woods in Northampton shire; also neere Colchester, and in many places alongst the high way leading from London to Henningham, in the county of Essex. The male Linden tree groweth in my Lord Treasurers garden in the Strand, and in sundry other places, as at Barn-elmes, and in a garden at Saint Katherines neere London.

The leaves of *Tilia* boyled in Smithes water with a

piece of Allum and a little honey, cure the sores in childrens mouthes.

The floures are commended by divers against paine of the head proceeding of a cold cause, against dissinesse, the Apoplexie, and also the falling sicknesse, and not onely the floures, but the distilled water thereof.

The leaves of the Linden (saith *Theophrastus*) are very sweet, and be a fodder for most kind of cattle: the fruit can be eaten of none.

BIRCH TREE

The common Birch tree waxeth likewise a great tree, having many boughs beset with many small rods or twigs, very limber and pliant; the barke of the yong twigs and

branches is plain, smooth, and full of sap, in colour like the chestnut, but the rind of the body or trunk is hard without, white, rough, and uneven, full of chinks or crevises: under which is found another fine barke, plaine, smooth, and as thin as paper, which heretofore was used in stead of paper to write on, before the making of paper was knowne: in Russia and these cold countries it serveth in stead of tiles and slate to cover their houses withall. This tree beareth for his flours certaine aglets like the Hasel tree, but smaller, wherein the seed is contained.

Birch

92

The common Birch tree grows in woods, fenny grounds, and mountains, in most places of England.

The catkins or aglets do first appear, and then the leaves, in Aprill or a little later.

Concerning the medicinable use of the Birch tree, or his parts, there is nothing extant either in the old or new writers.

This tree, saith *Pliny, lib.* 16. *cap.* 18. *Mirabili candore & tenuitate terribilis magistratuum virgis*: for in times past the magistrats rods were made thereof; and in our time also Schoolmasters and Parents do terrifie their children with rods made of Birch.

It serveth well to the decking up of houses and banqueting rooms, for places of pleasure, and for beautifying of streets in the Crosse and Gangweeke, and such like.

APPLE TREE

The Apple tree hath a body or trunke commonly of a meane bignesse, not very high, having long armes or branches, and the same disordered: the barke somewhat plaine, and not very rugged: the leaves bee also broad, more long than round, and finely nicked in the edges. The floures are whitish tending unto a blush colour. The fruit or Apples do differ in greatnesse, forme, colour, and taste; some covered with a redde skinne, others yellow or greene, varying infinitely according to the soyle and climate, some very great, some little, and many of a middle sort; some are sweet of taste, or something soure; most be of a middle taste betweene sweet and soure, the which to distinguish I thinke it impossible; notwithstanding I heare of one that intendeth to write a peculiar volume of Apples, and the use of them; yet when he hath done what he can doe, hee hath done nothing touching their severall kindes to distinguish them. This that hath beene said shall suffice for our History.

The tame and graffed Apple trees are planted and set in gardens and orchards made for that purpose: they delight to grow in good and fertile grounds: Kent doth abound with Apples of most sorts. But I have seene in the pastures and hedge-rows about the grounds of a worship-full Gentleman dwelling two miles from Hereford, called Master *Roger Bodnome*, so many trees of all sorts, that the servants drinke for the most part no other drinke but that which is made of Apples. The quantity is such, that by the report of the gentleman himselfe, the Parson hath for tithe many hogsheads of Syder. The hogs are fed with the fallings of them, which are so many, that they make choise of those Apples they do eate, who will not taste of any but of the best. An example doubtlesse to be followed of Gentlemen that have land and living: but envie saith, the poore will breake downe our hedges, and wee shall have the least part of the fruit; but forward in the name of God, graffe, set, plant and nourish up trees in every corner of your ground, the labour is small, the cost is nothing, the commodity is great, your selves shall have plenty, the poore shall have somwhat in time of want to relieve their necessitie, and God shall reward your good mindes and diligence.

ROSES

The Rose doth deserve the chief and prime place among all floures whatsoever; beeing not onely esteemed for his beauty, vertues, and his fragrant and odoriferous smell; but also because it is the honor and ornament of our English Scepter, as by the conjunction appeareth, in the uniting of those two most Royall Houses of Lancaster and Yorke. Which pleasant floures deserve the chiefest place in crownes and garlands, as *Anacreon Thius* a most antient Greeke Poet affirmes in those Verses of a Rose beginning thus;

The Rose is the honour and beauty of floures,
The Rose in the care and love of the Spring:
The Rose is the pleasure of th' heavenly Pow'rs.
The Boy of faire *Venus*, *Cythera's* Darling,
Doth wrap his head round with garlands of Rose,
When to the dances of the Graces he goes.

Augerius Busbequius speaking of the estimation and honor of the Rose, reporteth, That the Turks can by no means endure to see the leaves of Roses fall to the ground, because some of them have dreamed, that the first or most antient Rose did spring out of the bloud of *Venus*: and others of the Mahumetans say that it sprang of the sweat of *Mahumet*.

The Province or Damaske Rose

But there are many kindes of Roses, differing either in the bignesse of the floures, or the plant it selfe, roughnesse or smoothnesse, or in the multitude or fewnesse of the flours, or else in colour and smell; for divers of them are high and tall, others short and low, some have five leaves, others very many. Moreover, some be red, others white, and most of them or all sweetly smelling, especially those of the garden.

If the Curious could so be content, one generall description might serve to distinguish the whole stock or kindred of the Roses, being things so wel knowne: notwithstanding I thinke it not amisse to say somthing of

95

them severally, in hope to satisfie all. The white Rose hath very long stalkes of a wooddy substance, set or armed with divers sharpe prickles: the branches wherof are likewise full of prickles, whereon grow leaves consisting of five leaves for the most part, set upon a middle rib by couples, the old leaf standing at the point of the same, and every one of those small leaves somwhat snipt about the edges, somewhat rough, and of an overworne greene colour: from the bosome whereof shoot forth long foot-stalks, whereon grow very faire double flours of a white colour, and very sweet smell, having in the middle a few yellow threds or chives; which being past, there succeedeth a long fruit, greene at the first, but red when it is ripe, and stuffed with a downy choking matter, wherein is contained seed as hard as stones. The root is long, tough, and of a wooddy substance.

The red Rose groweth very low in respect of the former: the stalks are shorter, smoother, and browner of colour: The leaves are like, yet of a worse dusty colour: The floures grow on the tops of the branches, consisting of many leaves of a perfect red colour: the fruit is likewise red when it is ripe: the root is wooddy.

The common Damaske Rose in stature, prickely branches, and in other respects is like the white Rose; the especiall difference consists in the colour and smell of the flours: for these are of a pale red colour, of a more pleasant smel, and fitter for meat and medicine.

The *Rosa Provincialis minor* or lesser Province Rose differeth not from the former, but is altogether lesser: the floures and fruit are like: the use in physicke also agreeth with the precedent.

The Rose without prickles hath many young shoots comming from the root, dividing themselves into divers branches, tough, and of a wooddy substance as are all the rest of the Roses, of the height of two or three cubits, smooth and plain without any roughnesse or prickles at all: whereon grow leaves like those of the Holland Rose, of a shining deep green colour on the upper side, underneath somewhat hoary and hairy. The flours grow at the tops of the branches, consisting of an infinite number of leaves, greater than those of the Damaske Rose, more double, and of a colour between the red and damask Roses, of a most sweet smell. The fruit is round, red when it is ripe, and stuffed with the like flocks and seeds of those of the damaske Rose. The root is great, wooddy, and far spreading.

The Holland or Province Rose hath divers shoots proceeding from a wooddy root ful of sharpe prickles, dividing it selfe into divers branches, wheron grow leaves consisting of five leaves set on a rough middle rib, & those snipt about the edges: the flours grow on the tops of the branches, in shape and colour like the damaske Rose, but greater and more double, insomuch that the yellow chives in the middle are hard to be seene; of a reasonable good smell, but not fully so sweet as the common damaske Rose: the fruit is like the other of his kinde.

All these sorts of Roses we have in our London gardens, except that Rose without pricks, which as yet is a stranger in England. The double white Rose groweth wilde in many hedges of Lancashire in great aboundance, even as Briers do with us in these Southerly parts.

These floure from the end of May to the end of August, and divers times after, by reason the tops and superfluous branches are cut away in the end of their flouring: and then doe they somtimes floure even untill October and after.

The distilled water of Roses is good for the strengthning of the heart, and refreshing of the spirits, and likewise for all things that require a gentle cooling. The same being put in junketting dishes, cakes, sauces, and many other pleasant things, giveth a fine and delectable taste.

It mitigateth the paine of the eies proceeding of a hot cause, bringeth sleep, which also the fresh roses themselves provoke through their sweet and pleasant smell.

Of like vertue also are the leaves of these preserved in Sugar, especially if they be onely bruised with the hands, and diligently tempered with Sugar, and so heat at the fire rather than boyled.

The conserve of Roses, as well that which is crude and raw, as that which is made by ebullition or boiling, taken in the morning fasting, and last at night, strengthneth the heart, and taketh away the shaking and trembling thereof, and in a word is the most familiar thing to be used for the purposes aforesaid, and is thus made:

Take Roses at your pleasure, put them to boyle in faire water, having regard to the quantity; for if you have many Roses you may take more water; if fewer, the lesse water will serve: the which you shall boyle at the least three or foure houres, even as you would boile a piece of meate, untill in the eating they be very tender, at which time the Roses will lose their colour, that you would thinke your labour lost, and the thing spoiled. But proceed, for though the Roses have lost their colour, the water hath gotten the tincture thereof; then shall you adde unto one pound of Roses, foure pound of fine sugar in pure pouder, and so according to the rest of the Roses. Thus shall you let them boyle gently after the sugar is put therto, continually stirring it with a woodden Spatula untill it be cold, whereof one pound weight is worth six pound of the crude or raw conserve, as well for the vertues and goodnesse in taste, as also for the beautifull colour.

The making of the crude or raw conserve is very well knowne, as also Sugar roset, and divers other pretty things made of Roses and Sugar, which are impertent unto our history, because I intend nether to make thereof an Apothecaries shop, nor a Sugar-Bakers storehouse, leaving the rest for our cunning confectioners.

M U S K E   R O S E S

There be divers sorts of Roses planted in gardens, besides those written of in the former chapter, which are of most writers reckoned among the wilde Roses, notwithstanding we think it convenient to put them into a chapter betweene those of the garden and the brier Roses, as indifferent whether to make them of the wilde Roses, or of the tame, seeing we have made them denizons in our gardens for divers respects, and that worthily.

The single Muske Rose hath divers long shoots of a greenish colour and wooddy substance, armed with very sharpe prickles, dividing it selfe into divers branches : whereon doe grow long leaves, smooth and shining, made of divers leaves set upon a middle rib, like the other Roses: the floures grow on the tops of the branches, of a white colour, and pleasant

The Yellow Rose

sweet smell, like that of Muske, whereof it tooke his name; having certaine yellow seeds in the middle, as the rest of the Roses have: the fruit is red when it is ripe, and filled with such chaffie flockes and seeds as those of the other Roses: the root is tough and wooddy.

The yellow Rose (as divers do report) was by Art so coloured, and altered from his first estate, by grafting a wilde Rose upon a Broome-stalke; whereby (say they) it

99

doth not onely change his colour, but his smell and force. But for my part I having found the contrary by mine owne experience, cannot be induced to beleeve the report: for the roots and off-springs of this Rose have brought forth yellow roses, such as the maine stock or mother bringeth out, which event is not to be seen in all other plants that have been graffed. Moreover, the seeds of yellow roses have brought forth yellow Roses, such as the floure was from whence they were taken; which they should not do by any conjecturall reason, if that of themselves they were not a naturall kinde of Rose. Lastly it were contrary to that true principle,

Naturæ sequitur femina quodque suæ: that is to say.

Every seed and plant bringeth forth fruit like unto it selfe, both in shape and nature: but leaving that errour, I will proceed to the description: the yellow rose hath browne and prickly stalks or shoots, five or six cubits high, garnished with many leaves, like unto the Muske rose, of an excellent sweet smell, and more pleasant than the leaves of the Eglantine: the floures come forth among the leaves, and at the top of the branches of a faire gold yellow colour: the thrums in the middle, are also yellow: which being gone, there follow such knops or heads as the other Roses do beare.

The Canell or Cinnamon Rose, or the Rose smelling like Cinnamon, hath shots of a brown colour, foure cubits high, beset with thorny prickles, and leaves like unto those of Eglantine, but smaller and greener, of the savour or smell of Cinnamon, whereof it tooke his name, and not of the smell of his floures (as some have deemed) which have little or no savour at all: the floures be exceeding double, and yellow in the middle, of a pale red colour, and sometimes of a carnation: the root is of a wooddy substance.

These Roses are planted in our London Gardens, and else-where, but not found wilde in England.

The Muske Rose floureth in Autumne, or the fall of the leafe: the rest floure when the Damask and red Rose do.

The first is called *Rosa Moschata*, of the smell of Muske, as we have said: in Italian, *Rosa Moschetta*: in French, *Roses Musquees*, or *Muscadelles*: in low Dutch, *Musket roosen*: in English Musk Rose: the Latine and English titles may serve for the rest.

The white leaves stamped in a wooden dish with a piece of Allum and the juyce strained forth into some glased vessell, dried in the shadow, and kept, is the most fine and pleasant yellow colour that may be divised, not onely to limne or wash pictures and Imagerie in books, but also to colour meats and sauces, which notwithstanding the Allum is very wholesome.

W I L D E   R O S E S

The sweet Brier doth oftentimes grow higher than all the kindes of Roses; the shoots of it are hard, thicke, and wooddy; the leaves are glittering, and of a beautifull greene colour, of smell most pleasant: the Roses are little, five leaved, most commonly whitish, seldom tending to purple, of little or no smell at all: the fruit is long, of colour somewhat red, like a little olive stone, & like the little heads or berries of the others, but lesser than those of the garden: in which is contained rough cotton, or hairy downe and seed, folded and wrapped up in the same, which is small and hard: there be likewise found about the slender shoots hereof, round, soft, and hairy spunges, which we call Brier Balls, such as grow about the prickles of the Dog-Rose.

We have in our London gardens another sweet Brier, having greater leaves, and much sweeter: the floures likewise are greater, and somewhat doubled, exceeding sweet of smell, wherein it differeth from the former.

The Brier Bush or Hep tree, is also called *Rosa canina*, which is a plant so common and well knowne, that it were to small purpose to use many words in the description thereof: for even children with great deligh eat the berries thereof when they be ripe, make chaines and other prettie gewgawes of the fruit: cookes and gentlewomen make Tarts and such like dishes for pleasure thereof, and therefore this shall suffice for the description.

The Pimpinell Rose is likewise one of the wilde ones, whose leaves, consisting of divers small ones, are set upon a middle rib like those of Burnet, whereupon it was called the Burnet Rose. It growes very plentifully in a field as you go from a village in Essex, called Graies (upon the brinke of the river Thames) unto Horndon on the hill, insomuch that the field is full fraught therewith all over. It groweth likewise in a pasture as you goe from a village hard by London called Knights brige unto Fulham, a Village thereby.

The fruit when it is ripe maketh most pleasant meats and banqueting dishes, as tarts and such like; the making whereof I commit to the cunning cooke, and teeth to eate them in the rich mans mouth.

Larks heele or Larks claw

The garden Larks spur hath a round stem ful of branches, set with tender jagged leaves: the floures grow alongst the stalks toward the tops of the branches, of a blew colour, consisting of five little leaves which grow together and make one hollow floure, having a taile or spur at the end turning in like the spur of Todeflax. After come the seed, very blacke, like those of Leekes: the root perisheth at the first approch of Winter.

The second Larks spur is like the precedent, but somewhat smaller in stalkes and leaves: the floures are also like in forme, but of a white colour, wherein especially is the difference. These floures are sometimes of a purple colour, sometimes white, murrey, carnation, and of sundry other colours, varying infinitely, according to the soile or country wherein they live.

The wilde Larks spur hath most fine jagged leaves, cut and backt into divers parts, confusedly set upon a small middle tendrell: among which grow the floures, in shape like the others, but a great deale lesser, sometimes purple, otherwhiles white, and often of a mixt colour. The root is small and threddy.

These plants are set and sowne in gardens: the last groweth wilde in corne fields, and where corn hath grown. They floure for the most part all Summer long, from June to the end of August, and oft-times after.

Larks heele is called *Flos Regius*: of divers, *Consolida regalis*: who make it one of the Consounds or Comfreyes. It is

White or red Larks spur

also thought to be the *Delphinium* which *Dioscorides* describes in his third booke; wherewith it may agree: for the floures, and especially before they be perfected, have a certaine shew and likenesse of those Dolphins, which old pictures and armes of certain antient families have expressed with a crooked and bending figure or shape; by which signe also the heavenly Dolphine is set forth.

We finde little extant of the vertues of Larks heele, either in the antient or later writers, worth the noting, or

103

to be credited; yet it is set downe, that the seed of Larks spur drunken is good against the stingings of Scorpions; whose vertues are so forceable, that the herbe onely throwne before the Scorpion or any other venomous beast, couseth them to be without force or strength to hurt, insomuch that they cannot move or stirre untill the herbe be taken away: with many other such trifling toyes not worth the reading.

ROSE CAMPION

The first kind of Rose Campion hath round stalks very knotty and woolly, and at every knot or joint there do stand two woolly soft leaves like Mullein, but lesser & much narrower: the floures grow at the top of the stalke, of a perfect red colour.

The second Rose Campion differs not from the precedent in stalks, leaves, or fashion of the floures; the only difference consists in the colour, for the floures of this plant are of a milke white colour, and the other red.

The Rose Campion groweth plentifully in most gardens. Because the leaves thereof be soft, & fit to make weeks for candles, according to the testimony of *Dioscorides*, it was called *Lychnis*, that is, a Torch or such like light, according to the signification of the word, cleere, bright, and light-giving floures: and therefore they were called the Gardeners Delight, or the Gardeners Eye.

GARDEN POPPIES

The leaves of white Poppie are long, broad, smooth, longer than the leaves of Lettuce, whiter, and cut in the edges: the stem or stalke is straight and brittle, oftentimes a yard and a halfe high: on the top whereof grow white floures, in which at the very beginning appeareth a small head, accompanied with a number of threds or chives, which being full growne is round, and yet something long withall, and hath a cover or crownet upon the top; it is with many filmes or thin skin divided into coffers or severall partitions, in which is contained abundance of small round and whitish seed. The root groweth deepe, and is of no estimation nor continuance.

Like unto this is the blacke garden Poppy, saving that the floures are not so white and shining, but usually red, or at least spotted or straked with some lines of purple. The leaves are greater, more jagged, and sharper pointed. The seed is likewise blacker.

There are divers varieties of double Poppies, and their colours are commonly either white, red, darke purple, scarlet, or mixt of some of these.

These kinds of Poppies are sowne in gardens, and do afterward come of the fallings of their seed. The seed is perfected in July and August. This seed, as *Galen* saith in his booke of the faculties of Nourishments, is good to season bread with, but the white is better than the black. He also addeth, that the same is cold and causeth sleep, and yeeldeth no commendable nourishment to the body: it is often used in comfits, served at the table with other junketting dishes.

The oile which is pressed out of it is pleasant and delightfull to be eaten, & is taken with bread or any other waies with meat, without any sence of cooling.

A greater force is in the knobs or heads, which do specially prevaile to move sleepe, and to stay and represse distillations or rheums, and come neere in force to *Opium*, but more gentle. *Opium*, or the condensed juice of Poppy heads, is strongest of all; *Meconium* (which is the juice of the heads and leaves) is weaker. Both of them any waies taken either inwardly, or outwardly applied to the head, provoke sleepe. *Opium* somewhat too plentifully taken doth also bring death.

105

It mitigateth all kinds of paines; but it leaveth behinde it oftentimes a mischiefe worse than the disease it selfe, and that hard to be cured, as a dead palsie and such like.

So also collyries or eye medicines made with *Opium* have bin hurtfull to many; insomuch that they have weakened the eies and dulled the sight of those that have used it: whatsoever is compounded of *Opium* to mitigate the extreame paines of the eares, bringeth hardnesse of hearing. Wherefore all those medicines and compounds are to bee shunned that are to be made of *Opium*, and are not to be used but in extreame necessitie; and that is, when no other mitigater or assuager of paine doth any thing prevaile.

The leaves of Poppie boiled in water with a little sugar and drunke, cause sleep: or if it be boiled without sugar, and the head, feet, and temples bathed therewith, it doth effect the same.

The heads of Poppie boiled in water with sugar to a syrrup cause sleepe, and are good against rheumes and catarrhes that distil and fall down from the brain into the lungs, and ease the cough.

CORNE-ROSE OR WILDE POPPY

The stalks of red Poppy be blacke, tender, and brittle, somewhat hairy: the leaves are cut round about with deepe gashes like those of Succorie or wild Rocket. The flours grow forth at the tops of the stalks, being of a beautifull and gallant red colour, with blackish threds compassing about the middle part of the head, which being fully growne, is lesser than that of the garden Poppy: the seed is small and blacke.

The fields are garnished and overspred with these wilde Poppies in June and August.

Most men being led rather by false experiments than

reason, commend the floures against the Pleurisie, giving to drinke as soon as the pain comes, either the distilled water, or syrrup made by often infusing the leaves. And yet many times it happens, that the paine ceaseth by that meanes, though hardly sometimes.

GARDEN FLAXE

Flax riseth up with slender and round stalks. The leaves thereof bee long, narrow, and sharpe pointed: on the tops of the sprigs are faire blew floures, after which spring up little round knobs or buttons, in which is contained the seed, in forme somewhat long, smooth, glib or slipperie, of a darke colour.

Wild Poppy

Pliny saith that it is to be sowne in gravelly places, especially in furrowes: and that it burneth the ground, and maketh it worser: which thing also *Virgil* testifieth in his Georgickes. In English thus :

> Flaxe and Otes sowne consume
> The moisture of a fertile field:
> The same worketh Poppy, whose
> Juyce a deadly sleepe doth yeeld.

Flaxe is sowne in the spring, it floureth in June and July. After it is cut downe (as *Pliny*, *lib.* 19. *cap.* 1. saith) the stalks are put into the water, subject to the heat of

107

the Sun, & some weight laid on them to be steeped therein; the loosenes of the rinde is a signe when it is well steeped: then is it taken up and dried in the Sun, and after used as most huswives can tell better than my selfe.

The oile which is pressed out of the seed, is profitable for many purposes in Physicke and Surgerie; and is used of painters, picture makers, and other artificers.

The seeds stamped with the roots of wild Cucumbers, draweth forth splinters, thornes, broken bones, or any other thing fixed in any part of the body.

FOX-GLOVES

Purple Fox-gloves

Fox-glove with the purple floure is most common; the leaves whereof are long, nicked in the edges, of a light greene, in manner like those of Mullein, but lesser, and not so downy: the stalke is straight, from the middle whereof to the top stand the floures, set in a course one by another upon one side of the stalke, hanging downwards with the bottome upward, in forme long, like almost to finger stalkes, whereof it tooke his name *Digitalis*, of a red purple colour, with certaine white spots dasht within the floure; after which come up round heads, in which lies the seed somewhat browne, and as small as that of Time. The roots are many slender strings.

The Fox-Glove with white floures differs not from the precedent but in the colour of the floures; for as the other were purple, these contrariwise are of a milke-white colour.

We have in our Gardens another sort hereof, which bringeth forth most pleasant yellow floures, and somewhat lesser than the common kinde, wherein they differ.

Fox-glove groweth in barren sandy grounds, and under hedges almost every where.

Fox-gloves some call in French, *Gantes nostre dame.*

The Fox-gloves in that they are bitter, are hot and dry, with a certaine kinde of clensing qualitie joyned therewith; yet are they of no use, neither have they any place amongst medicines, according to the Antients.

W O L F E S - B A N E

Aconite, of some called Thora (others adde thereto the place where it groweth in great aboundance, which is the Alps, and call it *Thora Valdensium*), tooke his name of the Greeke word signifying corruption, poison, or death, which are the certaine effects of this pernitious plant: for this they use very much in poison, and when they mean to infect their arrow heads, the more speedily and deadly to dispatch the wilde beasts which greatly annoy those Mountaines of the Alpes. To which purpose also it is brought into the Mart townes neere those places, to be sold unto the hunters, the juyce thereof being prepared by pressing forth, and so kept in hornes and hoofes of beast for the most speedy poyson of the Aconites: for an arrow touched therewith leaves the wound uncurable (if it but fetch bloud where it entred in) unlesse that round about the wound the flesh bee speedily cut away in great quantitie: this plant therefore may rightly be accounted as first and chiefe of those called Sagittaries or Aconites, by reason of the malignant qualities aforesaid. This that

hath beene sayd, argueth also that *Matthiolus* hath unproperly called it *Pseudoaconitum*, that is, false or bastard
Aconite; for without question there is no worse or
more speedie venome in the world, nor no Aconite or
toxible plant comparable hereunto. And yet let us consider the fatherly care and providence of God, who hath
provided a conquerour and triumpher over this plant
so venomous, namely his *Anti-
gonist*, *Antithora*, or to speake
in shorter and fewer syllables,
Anthora, which is the very antidote or remedie against this
kinde of Aconite.

The yellow kinde of Wolfes-
bane hath large shining green
leaves fashioned like a vine. His
stalks grow up to the height of
three cubits, bearing very fine
yellow floures, fantastically fashioned, and in such manner
shaped, that I can very hardly
describe them to you. This
plant groweth naturally in the
darke hilly forrests, & shadowie
woods, that are not travelled
nor haunted, but by wilde and

Broad leafed Wolfs-bane   savage beasts, and is thought
to bee the strongest and next unto *Thora* in his
poisoning qualitie, of all the rest of the Aconites, or
Woolfes banes; insomuch that if a few of the floures
be chewed in the mouth, and spit forth againe presently,
yet forthwith it burneth the jaws and tongue, causing
them to swell, and making a certain swimming or
giddinesse in the head. This calleth to my remembrance an history of a certain Gentleman dwelling in
Lincolneshire, called *Mahewe*, the true report whereof

my very good friend Mr. *Nicholas Belson,* somtimes
Fellow of Kings Colledge in Cambridge, hath delivered
unto me: Mr. *Mahewe* dwelling in Boston, a student in
physick, having occasion to ride through the fens of
Lincolnshire, found a root that the hogs had turned up,
which seemed unto him very strange and unknowne, for
that it was in the spring before the leaves were out: this
he tasted, and it so inflamed his mouth, tongue, and lips,
that it caused them to swell very extremely, so that before
he could get to the towne of Boston, he could not speake,
and no doubt had lost his life if that the Lord God had
not blessed those good remedies which presently he procured and used. I have here thought good to expresse
this history, for two speciall causes; the first is, that some
industrious and diligent observer of nature may be provoked to seeke forth that venomous plant, or some of
his kindes: for I am certainly persuaded that it is either
the *Thora Valdensium,* or *Aconitum luteum,* whereof this
gentleman tasted, which two plants have not at any
time bin thought to grow naturally in England: the
other cause is, for that I would warne others to beware
by that gentlemans harme.

MONKES HOOD

Helmet-floure, or the great Monkes-hood, beareth very
faire and goodly blew floures in shape like an Helmet;
which are so beautifull, that a man would thinke they
were of some excellent vertue, but *non est semper fides
habenda fronti.* This plant is universally knowne in our
London gardens and elsewhere; but naturally it groweth
in the mountaines of Rhetia, and in sundry places of the
Alps, where you shall find the grasse that groweth round
it eaten up with cattell, but no part of the herbe it selfe
touched, except by certaine flies, who in such abundant

measure swarme about the same that they cover the whole plant: and (which is very straunge) although these  flies do with great delight feed hereupon, yet of them there is confected an Antidot or most availeable medicine against the deadly bite of the spider called *Tarantala*, or any other venomous beast whatsoever; yea, an excellent remedy not only against the Aconites, but all other poisons whatsoever. The medicine of the foresaid flies is thus made: Take of the the flies which have fed themselves as is above mentioned, in number twentie, of *Aristolochia rotunda*, and bole Armoniack, of each a dram.

The force and facultie of this Wolfs-bane is deadly to man and all kindes of beasts: the same was tried of late

Monkes hood

in Antwerpe, and is as yet fresh in memorie, by an evident experiment, but most lamentable; for when the leaves hereof were by certaine ignorant persons served up in sallads, all that did eat thereof were presently taken with most cruell symptomes, and so died.

The symptomes that follow those that doe eat of these deadly Herbs are these; their lipps and tongue swell forthwith, their eyes hang out, their thighes are stiffe, and their wits are taken from them, as *Avicen* writes, *lib.* 4. The force of this poison is such, that if the points of darts or arrowes be touched therewith, it brings deadly hurt to those that are wounded with the same.

Against so deadly a poison *Avicen* reckoneth up certain remedies, which help after the poyson is vomited up: and among these he maketh mention of the Mouse (as the copies every where have it) nourished and fed up with *Napellus,* which is altogether an enemie to the poisonsome nature of it, and delivereth him that hath taken it from all perill and danger.

Antonius Guanerius of Pavia, a famous physition in his age, in his treaty of poisons is of opinion, that it is not a mouse which *Avicen* speaketh of, but a fly: for he telleth of a certaine Philosopher who did very carefully and diligently make search after this mouse, and neither could find at any time any mouse, nor the roots of Wolfsbane gnawn or bitten, as he had read: but in searching he found many flies feeding on the leaves, which the said Philosopher tooke, and made of them an antidote or counterpoison, which hee found to be good and effectuall against other poisons, but especially against the poison of Wolfs-bane.

The composition consisteth of two ounces of *Terra lemnia,* as many of the berries of the Bay tree, and the like weight of Mithridate, 24 of the flies that have taken their repast upon Wolfes-bane, of hony and oile Olive a sufficient quantitie.

The same opinion that *Guanerius* is of, *Pena* and *Lobel* do also hold; who affirme, that there was never seene at any time any mouse feeding thereon, but that there bee flies which resort unto it by swarmes, and feed not only upon the floures, but on the herb also.

There hath bin little heretofore set down concerning the Vertues of Aconites, but much might be said of the hurts that have come hereby, as the wofull experience as the lamentable example at Antwerp yet fresh in memorie, doth declare, as we have said.

113

MITHRIDATE WOLFES-BANE

This plant called *Anthora*, being the antidote against the poison of *Thora*, *Aconite* or Wolfes bane, hath slender hollow stalkes, very brittle, a cubit high, garnished with fine cut or jagged leaves, very like to *Nigella Romana*, or the common Larks spurre, called *Consolida regalis*: at the top of the stalks grow faire floures, fashioned like a little helmet, of an overworne yellow colour; after which come small blackish cods, wherein is contained blacke shining seed like those of Onions: the root consisteth of divers knobs or tuberous lumps, of the bignesse of a mans thumbe.

This plant groweth abundantly in the Alpes, called *Rhetici*, in Savoy, and in Liguria. The Ligurians of Turnin, and those that dwell neer the lake Lemane, have found this herbe to be a present remedie against the deadly poison of the herb *Thora*, and the rest of the Aconits, provided that when it is brought into the garden there to be kept for phisicks use, it must not be planted neere to any of the Aconites: for through his attractive qualitie, it will draw unto it selfe the maligne and venomous poison of the Aconite, whereby it wil become of the like qualitie, that is, to become poisonous likewise: but being kept far off, it retaineth his owne naturall qualitie still. The root is wonderfull bitter, it killeth and driveth forth all manner of wormes of the belly.

It is called *Anthora*, as though they should say *Antithora*, because it is an enemy to *Thora*, and a counterpoyson to the same. *Thora* and *Anthora*, or *Tura* and *Antura*, seem to be new words, but yet they are used in *Marcellus Empericus*, an old writer, who teaches us a medicine to be made of *Tura* and *Antura*, against the pin and web in the eies: in English, yellow Monks-hood, yellow Helmet floure, and Aconites Mithridate.

SEA LAVANDER

There hath beene among writers from time to time great contention about this plant *Limonium*, no one Author agreeing with another: for some have called this herbe *Limonium*; some another herbe by this name; and some in remooving the rocke, have mired themselves in the mud, as *Matthiolus*, who described two kindes, but made no distinction of them, nor yet expressed which was the true *Limonium*; but as a man herein ignorant, he speakes not a word of them. Now then to leave controversies and cavilling, the true *Limonium* is that which hath faire leaves, like the Limon or Orenge tree, but of a darke greene colour, somewhat fatter, and a little crumpled: amongst which leaves riseth up an hard and brittle naked stalke of a foot high, divided at the top into sundry other small branches, which grow for the most part upon one side, full of little blewish floures, in shew like Lavander, with long red seed, and a thicke root like unto the small Docke.

Sea Lavander

There is a kinde of *Limonium* like the first in each respect, but lesser, which groweth upon rockes and chalkie cliffes.

The first groweth in great plenty upon the walls of the fort against Gravesend: but abundantly on the bankes of the River below the same towne, as also below the

115

Kings Store-house at Chattam: and fast by the Kings
Ferrey going into the Isle of Shepey: in the salt marshes
by Lee in Essex: in the Marsh by Harwich, and many
other places.

The small kind I could never find in any other place
but upon the chalky cliffe going from the towne of
Margate downe to the sea side, upon the left hand.

It shall be needlesse to trouble you with any other
Latine name than is exprest in their titles: The people
neere the sea side where it growes do call it Marsh Lavander, and sea Lavander.

SERAPIA'S TURBITH, OR SEA STAR-WORT

Tripolium hath long and large leaves somewhat hollow or
furrowed, of a shining green colour declining to blewnesse: among which riseth up a stalke of two cubits high
and more, which toward the top is divided into many
small branches garnished with many floures like Camomill, yellow in the middle, set about or bordered with
small blewish leaves like a pale; which grow into a
whitish rough downe that flieth away with the wind.
The root is long and threddy.

These herbs grow plentifully alongst the English
coasts in many places, as by the fort against Gravesend,
in the Isle of Shepey in sundry places, in a marsh which
is under the town wals of Harwich, in the marsh by Lee
in Essex, in a marsh which is between the Isle of Shepey
and Sandwich, especially where it ebbeth and floweth:
being brought into gardens it flourisheth a long time,
but there it waxeth huge, great, and ranke, and changeth
the great roots into strings.

It is reported by men of great fame and learning, That
this plant was called *Tripolium* because it doth change
the colour of his floures thrice in a day. This rumor we

may beleeve as true, for that we see and perceive things of as great or greater wonder to proceed out of the earth. This herbe I planted in my garden, whither in his season I did repaire to finde out the truth hereof, but I could not espy any such variablenesse herein: yet thus much I may say, that as the heate of the sun doth change the colour of divers floures, so it fell out with this, which in the morning was very faire, but afterward of a pale or wan colour. Which proveth that to be but a fable which *Dioscorides* saith is reported by some, that in one day it changeth the colour of his floures thrice; that is to say, in the morning it is white, at noone purple, and in the evening crimson. But it is not untrue, that there may be found three colours of the floures in one day, by reason that the floures are not all perfected together, (as before I partly touched) but one after another by little and little. And there may easily be observed three colours in them, which is to be understood of them that are beginning to floure, that are perfectly floured, and those that are falling away. For they that are blowing and be not wide open and perfect are of a purplish colour, and those that are perfect and wide open of a whitish blew, and such as have fallen away have a white down: which changing hapneth unto sundry other plants. This herbe is called of *Serapio, Turbith*: women that dwell by the sea side call it in English, blew Daisies, or blew Camomill; & about Harwich it is called Hogs beans, for that the swine do greatly desire to feed thereon, as also for that the knobs about the roots doe somewhat resemble the garden bean.

LAND PLANTAINE

As the Greeks have called some kinds of herbs Serpents tongue, Dogs tongue, and Ox tongue; so have they termed a kinde of Plantain *Arnoglosson*, which is as if you should say Lambs tongue, well known to all, by reason of the great commoditie and plenty of it growing every where; and therefore it is needlesse to spend time about them. The greatnes and fashion of the leaves hath been the cause of the varieties and diversities of their names.

Plantaine

The second is like the first, and differeth in that, that this Plantaine hath greater but shorter spikes or knaps; and the leaves are of an hoary or overworne green colour: the stalks are likewise hoary and hairy.

The juice dropped in the eies cooles the heate and inflammation thereof. I find in antient writers many good-morrowes, which I thinke not meet to bring into your memorie againe; as, That three roots will cure one griefe, foure another disease, six hanged about the necke are good for another malady, &c. all which are but ridiculous toyes.

Spurge

The first kinde of Sea Spurge riseth forth of the sands, or baich of the sea, with sundry reddish stems or stalkes growing upon one single root; and the stalkes are beset with small, fat, and narrow leaves like unto the leaves of Flax. The floures are yellowish, and grow out of little dishes or Saucers like the common kinde of Spurge. After the floures come triangle seeds, as in the other Tithymales.

The second kinde (called *Helioscopius*, or *Solisequius*: and in English, according to his Greeke name, Sunne Spurge, or time Tithymale, of turning or keeping time with the Sunne) hath sundry reddish stalkes of a foot high: the leaves are like unto Purslane, not so great nor thicke, but snipt about the edges: the floures are yellowish, and growing in little platters.

Spurge

The first kinde of Spurge groweth by the sea side upon the rowling Sand and Baich, as at Lee in Essex, at Langtree point right against Harwich, at Whitstable in Kent, and in many other places.

The second groweth in grounds that lie waste, and in barren earable soile, almost every where.

First the milke and sap is in speciall use, then the fruit and leaves, but the root is of least strength. The strongest kinde of Tithymale, and of greatest force is that of the sea.

Some write by report of others, that it enflameth exceedingly, but my selfe speak by experience; for walking along the sea coast at Lee in Essex, with a Gentleman called Mr. *Rich*, dwelling in the same towne, I tooke but one drop of it into my mouth; which neverthelesse did so inflame and swell in my throte that I hardly escaped with my life. And in like case was the Gentleman, which caused us to take our horses, and poste for our lives unto the next farme house to drinke some milke to quench the extremitie of our heat, which then ceased.

119

The juyce mixed with hony, causeth haire to fall from that place which is anointed therewith, if it be done in the Sun.

The juyce or milke is good to stop hollow teeth, being put into them warily, so that you touch neither the gums, nor any of the other teeth in the mouth with the said medicine. The same cureth all roughnesse of the skin, and the white scurfe of the head. It killeth fish, being mixed with any thing that they will eat.

These herbes by mine advise would not be received into the body, considering that there be so many other good and wholesome potions to be made with other herbes, that may be taken without perill.

NAVELWOORT, OR PENNIWOORT OF THE WALL

The great Navelwoort hath round and thicke leaves, somewhat bluntly indented about the edges, and somewhat hollow in the midst on the upper part, having a short tender stemme fastened to the middest of the leafe, on the lower side underneath the stalke, whereon the floures do grow, is small and hollow, an handfull high and more, beset with many small floures of an overworne incarnate colour. The root is small like an olive, of a white colour.

The second kinde of Wall Penniwort or Navelwoort hath broad thicke leaves somewhat deepely indented about the edges: spred upon the ground in manner of a tuft, set about the tender stalke; among which riseth up a tender stalke whereon doe grow the like leaves. The floures stand on the top consisting of five small leaves of a whitish colour, with redde spots in them.

There is a kinde of Navelwoort that groweth in watery places, which is called of the husbandman Sheeps bane, because it killeth sheepe that do eat thereof: it is not much unlike the precedent, but the round edges of the leaves are not so even as the other; and this creepeth upon the ground, and the other upon the stone walls.

The first kind of Penniwoort groweth plentifully in Northampton upon every stone wall about the towne, at Bristow, Bathe, Wells, and most places of the West countrie upon stone walls. It groweth upon Westminster Abbey, over the doore that leadeth from *Chaucers* tombe to the old palace.

The second and third grow upon the Alpes neere Piedmont, and Bavier, and upon the mountaines of Germany: I found the third growing upon Bieston Castle in Cheshire.

Navelwoort is called of some, *Hortus Veneris*, or Venus garden, and *Terræ umbilicus*, or the Navel of the earth: in English, Penniwoort, Wall-Penniwoort, Ladies Navell, Hipwoort and Kidney-woort. Water Penniwoort is called in English, Sheepe-killing Pennigrasse.

The ignorant Apothecaries doe use the Water Penny-wort in stead of this of the wall, which they cannot doe without great error, and much danger to the patient: for husbandmen know well, that it is noisome unto Sheepe, and other cattell that feed thereon, and for the most part bringeth death unto them, much more to men by a stronger reason.

SAMPIER

Rocke Sampier hath many fat and thicke leaves somwhat like those of the lesser Purslane, of a spicie taste, with a certain saltnesse; amongst which rises up a stalk divided into many smal spraies or sprigs, on the top whereof grow spoky tufts of white floures, like the tufts of Fennell or Dill; after that comes the seed, like the seed of Fenell, but greater: the root is thicke and knobby, beeing of smell delightfull and pleasant.

121

The second Sampier, called *Pastinaca marina*, or sea Parsnep, hath long fat leaves very much jagged or cut even to the middle rib, sharp or prickely pointed, which are set upon large fat jointed stalks; on the top wherof do grow tufts of whitish or else reddish floures. The seed is wrapped in thorny husks: the root is thicke and long, not unlike to the Parsenep, very good and wholsome to be eaten.

Sea Parsnep

Rocke Sampier growes on the rocky clifts at Dover, Winchelsey, by Rie, about Southampton, the Isle of Wight, and most rocks about the West and North parts of England.

The second groweth neere the sea upon the sands and Baych betweene Whitstable and the Isle of Tenet, by Sandwich, and by the sea neere West-chester.

The leaves kept in pickle, and eaten in sallads with oile and vineger, is a pleasant sauce for meat.

GLASSE SALTWORT

Glassewort hath many grosse thicke and round stalks a foot high, full of fat and thicke sprigs, set with many knots or joints, without any leaves at all, of a reddish greene colour: the whole plant resembles a branch of Corall. These plants are to be found in salt marshes almost everie where.

Stones are beaten to pouder and mixed with ashes, which beeing melted together, become the matter whereof glasse is made. Which while it is made red hot in the furnace, and is melted, becomming liquid and fit to worke upon, doth yeeld as it were a fat floting aloft; which when it is cold waxeth as hard as a stone, yet is it brittle and quickly broken.

A great quantitie taken is mischievous and deadly: the smel and smoke also of this herb being burnt drives away serpents.

S. Johns Wort

Saint Johns wort hath brownish stalks beset with many small and narrow leaves, which if you behold betwixt your eies and the light, do appeare as it were bored or thrust thorow in an infinite number of places with pinnes points. The branches divide themselves into sundry smal twigs at the top whereof grow many yellow floures, which with the leaves bruised do yeeld a reddish juice of the colour of bloud. The seed is contained in little sharp pointed huskes blacke of colour, and smelling like Rosin. The root is long, yellow, and of a wooddy substance.

They grow very plentifully in pastures in every countrie.

S. Johns wort is called in Latine *Hypericum*: in shops, *Perforata*: of divers, *Fuga dæmonum*: in French *Mille Pertuys*: in English, S. Johns wort, or S. Johns grases.

The leaves, floures, and seeds stamped, and put into a glasse with oile olive, and set in the hot sun for certain weeks together, and then strained from those herbs, and the like quantitie of new put in and sunned in like manner, doth make an oile of the colour of bloud, which is a most pretious remedie for deep wounds and those that are thorow the body, for the sinues that are prickt, or any wound made with a venomed weapon. I am accustomed to make a compound oile hereof, the making of which you shall receive at my hands, because I know that in the world

S. Johns Wort

there is not a better, no not the naturall Balsam it selfe; for I dare undertake to cure any such wound as absolutely in each respect, if not sooner and better, as any man shall or may with naturall Balsam.

Take white wine two pintes, oile olive foure pounds, oile of Turpentine two pounds, the leaves, floures, and seeds of S. Johns wort of each two great handfulls gently bruised; put them all together into a great double glasse, and set it in the Sun eight or ten daies; then boile them in the same glasse *per Balneum Mariæ*, that is, in a kettle of water, with some straw in the bottome, wherein the glasse must stand to boile: which done, strain the liquor from the herbs, and do as you did before, putting in the like quantitie of herbs, floures, and seeds, but not any more wine. Thus have you a great secret for the purposes afore-said.

HOUSLEEKE OR SENGREENE

Great Housleek or Sengreene (syrnamed tree Housleeke) bringeth forth a stalke a cubit high, somtimes higher, and often two; which is thick, hard, wooddy, tough, and that can hardly be broken, parted into divers branches, and

124

covered with a thick grosse bark, which in the lower part reserveth certaine prints or impressed markes of the leaves that are fallen away. The leaves are fat, well bodied, full of juice, an inch long and somewhat more, like little tongues, very curiously minced in the edges, standing upon the tops of the braunches, having in them the shape of an eye. The floures grow out of the branches, which are divided into many springs; which floures are slender, yellow, and spred like a star; in their places commeth up very fine seed, the springs withering away: the root is parted into many off-springs. This plant is alwaies greene, neither is it hurt by the cold in winter, growing in his native soile; whereupon it is named *Sempervivum*, or Sengreene.

Great Sengreene is found growing of it selfe on the tops of houses, old walls, and such like places, in very many provinces of the East, and of Greece, and also in the Islands of the Mediterranean sea, as in Creet, now called Candy, Rhodes, Zant, and others: neither is Spain without it; for (as *Clusius* witnesseth) it groweth in many places of Portingall; otherwise it is cherished in pots. In cold countries and such as lie Northward, as in both the Germanies, it neither groweth of it selfe, nor yet lasteth long, though it be carefully planted, and diligently looked unto, but through the extremitie of the weather and the over-much cold of winter it perisheth.

They take away the fire of burnings and scaldings, and being applied with barly meale dried, do take away the paine of the gout.

The juice of Housleeke, garden Nightshade, and the buds of Poplar boiled in *Axungia porci*, or hogs grease, make the most singular Populeon that ever was used in Surgerie.

The juice hereof taketh away cornes from the toes and feet, if they be washed and bathed therewith, and every day and night as it were emplaistered with the skin of the same Housleeke, which certainly taketh them away without incision or such like, as hath been experimented by my verie good friend Mr. *Nicholas Belson*, a man painfull and curious in searching forth the secrets of nature.

The decoction of Housleek or the juice thereof cooleth the inflammation of the eyes, being dropped thereinto, and the herb bruised and layd upon them.

Our Ladies Slipper

Our Ladies Slipper

Our Ladies Shoo or Slipper hath a thicke knobbed root from which riseth up a stiffe and hairy stalke a foot high, set by certaine spaces with faire broad leaves. At the top of the stalke groweth one single floure, seldome two, fashioned on the one side like an egge; on the other side it is open, empty, and hollow, and of the form of a shoo or slipper, whereof it tooke his name; of a yellow colour on the outside, and of a shining deep yellow on the inside. The middle part is compassed about with four leaves of a bright purple colour, often of a light red or obscure crimson, and sometimes yellow as in the middle part, which in shape is like an egge as aforesaid.

Ladies Slipper groweth upon the mountains of Germany, Hungary, and Poland. I have a plant thereof in my garden, which I received from Mr. *Garret* Apothecarie, my very good friend.

Touching the faculties of our Ladies shoo we have

nothing to write, it beeing not sufficiently known to the old writers, no nor to the new.

BELL-FLOURES

Coventry-Bells have broad leaves rough and hairy, not unlike to those of the Golden Buglosse, of a swart greene colour: among which do rise up stiffe hairie stalkes the second yeare after the sowing of the seed: which stalkes divide themselves into sundry branches, whereupon grow many faire and pleasant bell-floures, long, hollow, and cut on the brim with five sleight gashes, ending in five corners toward night, when the floure shutteth it selfe up as doe most of the Bell-floures: in the middle of the floures be three or foure whitish chives, as also much downie haire, such as is in the eares of a Dog, or such like beast. The whole floure is of a blew purple colour: which being past, there succeed great square or cornered seed-vessels, divided on the inside into divers cels or chambers, wherein do lie scatteringly many small browne flat seeds. The root is long and great like a Parsenep, garnished with many threddy strings, which perisheth when it hath perfected his seed, which is in the second yeare after his sowing, and recovereth it selfe againe by the falling of the seed.

They grow in woods, mountaines, and dark vallies, and under hedges among the bushes, especially about Coventry, where they grow very plentifully abroad in the fields, and are there called Coventry bells, and of some about London, Canterbury bells; but unproperly, for that there is another kinde of Bell-floure growing in Kent about Canterbury, which may more fitly be called Canterbury Bells, because they grow there more plentifully than in any other countrey. These pleasant Bell-floures wee have in our London gardens especially for the beauty of their floure, although they be kinds of Rampions, and the roots eaten as Rampions are.

127

They floure in June, July, and August; the seed waxeth ripe in the mean time; for these plants bring not forth their floures all at once, but when one floureth another seedeth.

Coventry bels are called in Latine *Viola mariana*: in English, *Mercuries* Violets, or Coventry Rapes, and of some, Mariets.

The root is not used in physicke, but only for a sallet root boiled and eaten, with oile, vineger, and pepper.

Valerian

VALERIAN, OR SETWALL

The tame or garden Valerian and likewise the Greeke Valerian are planted in gardens; the wilde ones are found in moist places hard to rivers sides, ditches, and watery pits; yet the greater of these is brought into gardens where it flourisheth, but the lesser hardly prospereth.

Generally the Valerians are called by one name, in Latine, *Valeriana*; in shoppes also *Phu*: in English, Valerian, Capons taile, and Setwall; but unproperly, for that name belongeth to *Zedoaria*, which is not Valerian.

The dry root is put into counterpoysons and medicines preservative against the pestilence: whereupon it hath been had (and is to this day among the poore people of our Northerne parts) in such veneration amongst them, that no broths, pottage or physicall meats are worth any thing, if Setwall were not at an end: whereupon some

upon some woman Poët or other hath made these verses.
      They that will have their heale,
      Must put Setwall in their keale.

CHERVILL

The leaves of Chervill are slender, and diversly cut,
something hairy, of a whitish greene: the stalkes be
short, slender, round, and
hollow within, which at the
first together with the leaves
are of a whitish green, but
tending to a red when the
seeds are ripe : the floures
be white, and grow upon
scattered tufts.

Great Chervill hath large
leaves deepely cut or jagged,
in shew very like unto Hemlockes, of a very good and
pleasant smell and taste like
unto Chervill, and something
hairy, which hath caused us
to call it sweet Chervill. The
great sweet Chervill groweth
in my garden, and in the
gardens of other men who
have bin diligent in these
matters.

Chervill

Chervill is used very much among the Dutch people in
a kinde of Loblolly or hotchpot which they do eat, called
Warmus.

The leaves of sweet Chervill are exceeding good, wholesome and pleasant among other sallad herbs, giving the
taste of Anise seed unto the rest.

The seeds eaten as a sallad whiles they are yet green

with oile, vineger, and pepper, exceed all other sallads by many degrees, both in pleasantnesse of taste, sweetnesse of smell, and wholsomnesse for the cold and feeble stomacke.

The roots are likewise most excellent in a sallad, if they be boiled and afterwards dressed as the cunning Cooke knoweth how better than my selfe: notwithstanding I use to eat them with oile and vineger, being first boiled; which is very good for old people that are dull and without courage: it rejoiceth and comforteth the heart, and increaseth their lust and strength.

Wilde Time of Candy

WILDE TIME

The first is our common creeping Time, which is so well knowne, that it needeth no description; yet this ye shall understand, that it beareth floures of a purple colour, as every body knoweth. Of which kinde I found another sort, with floures as white as snow, and have planted it in my garden, where it becommeth an herbe of great beauty.

This wilde Time that bringeth forth white floures differeth not from the other, but onely in the colour of the floures, whence it may be called *Serpillum vulgare flore albo*, White floured Wilde Time.

Wilde Time of Candy is like unto the other wild

130

Times, saving that his leaves are narrower and longer, and more in number at each joynt. The smell is more aromaticall than any of the others, wherein is the difference.

The first groweth upon barren hills and untoiled places: the second groweth in Gardens. The white kinde I found at South fleet in Kent. They floure from May to the end of Summer.

Wild Time is called in Latine, *Serpillum*, *à serpendo*, of creeping: in English, wilde Time, Puliall mountaine, Pella Mountaine, running Time, creeping Time, Mother of Time. *Ælianus* in his ninth booke of his sundry Histories seemeth to number wilde Time among the floures. *Dionysius Junior* (saith he) comming into the city Locris in Italy, possessed most of the houses of the city, and did strew them with roses, wilde Time, and other such kindes of floures. Yet *Virgil* in the second Eclog of his Bucolicks doth most manifestly testifie, that wilde Time is an herbe, in these words:

> *Thestilis* for mowers tyr'd with parching heate,
> Garlicke, wilde Time, strong smelling herbes
> doth beate.

Out of which place it may be gathered, that common wilde time is the true and right *Serpillum*, or wilde Time.

ELECAMPANE

Elecampane bringeth forth presently from the root great white leaves, sharpe pointed, almost like those of great Comfrey, but soft, and covered with a hairy downe, of a whitish greene colour, and are more white underneath, sleightly nicked in the edges: the stalke is a yard and a halfe long, about a finger thicke, not without downe, divided at the top into divers branches, upon the top of every sprig stand great floures broad and round, of which not only the long smal leaves that compasse round about are yellow, but also the middle ball or circle, which is filled up with an infinite number of threds, and at length is turned into fine downe; under which is slender and long seed: the root is uneven, thicke, and as much as a man may gripe, not long, oftentimes blackish without, white within, and full of substance, sweet of smell, and bitter of taste.

It groweth in medowes that are fat and fruitfull: it is also oftentimes found upon mountains, shadowie places, that be not altogether dry: it groweth plentifully in the fields on the left hand as you go from Dunstable to Puddlehill : also in an orchard as you go from Colbrooke to Ditton ferry, which is the way to Windsor, and in sundry other places, as at Lidde, and Folkestone, neere to Dover by the sea side.

The floures are in their bravery in June and July: the roots be gathered in Autumne, and oftentimes in Aprill and May.

Some report that this plant tooke the name *Helenium* of *Helena* wife to *Menalaus*, who had her hands full of it when *Paris* stole her away into Phrygia.

The root of this Elecampane is marvellous good for many things.   It is good for shortnesse of breath, and an old cough, and for such as cannot breathe unlesse they hold their neckes upright.

It is of great vertue both given in a looch, which is a medicine to be licked on, and likewise preserved, as also otherwise given to purge and void out thicke, tough, and clammy humours, which sticke in the chest and lungs.

The root of Elecampane is with good success mixed with counterpoisons: it is good for them that are bursten and troubled with cramps and convulsions.

ORCHIS

There be divers kindes of Fox-stones, differing very much in shape of their leaves, as also in floures: some have floures, wherein is to be seen the shape of sundry sorts of living creatures; some the shape and proportion of flies, in other gnats, some humble bees, others like unto honey Bees; some like Butter-flies, and other like Waspes that be dead; some yellow of colour, others white; some purple mixed with red, others of a brown overworne colour: the which severally to distinguish, as well those

Birds Nest

here set downe, as also those that offer themselves daily to our view and consideration, would require a particular volume; for there is not any plant which doth offer such varietie unto us as these, except the Tulipa's which go beyond all account: for that the most singular Simplest that ever was in these later ages, *Carolus Clusius* (who for his singular industry and knowledge herein is worthy triple honor) hath spent at the least 35 yeares, sowing the seeds of Tulipa's from yeare to yeare, and to this day he could never attain to the end or certainty of their several kinds of colours.

Butterfly Orchis or Satyrion beares next the root two very broad leaves like those of the Lilly, seldome three: the floures be white of colour, resembling the shape of a butterfly: the stalke is a foot high.

The Waspe Satyrion groweth out of the ground, having stalks small and tender: the leaves are like the former, but somwhat greater, declining to a brown or dark colour. The flours be small, of the colour of a dry oken leafe, in shape resembling the great Bee called in English an Hornet, or drone Bee.

The leaves of Bee Satyrion are longer than the last before mentioned, narrower, turning themselves against the Sun as it were round. The stalk is round, tender, and very fragile. At the top grow the floures, resembling in shape the dead carkasse of a Bee. The bulbes of the roots be smaller and rounder than the last described.

Birds nest hath many tangling roots platted or crossed one over another very intricately, which resembleth a Crows nest made of sticks ; from which riseth up a thicke soft grosse stalk of a browne colour, set with small short leaves of the colour of a dry oken leafe that hath lien under the tree all the winter long. On the top of the stalke groweth a spiky eare or tuft of floures. This bastard or unkindely Satyrion is very seldome seene in these Southerly parts of England.

The other kindes of Orchis grow for the most part in moist medowes and fertile pastures, as also in moist woods. That kind which resembleth the white Butter-fly groweth upon the declining of the hill at the end of Hampsted heath, neere to a small cottage there in the way side, as yee goe from London to Henden a village there by. It groweth in the fields adjoyning to the fold or pin-fold without the gate, at a village called High-gate, neere London.

There is no great use of these in physicke, but they are chiefly regarded for the pleasant and beautifull floures wherewith Nature hath seemed to play and disport her selfe.

ROSEMARY

Rosemarie is a wooddy shrub, growing oftentimes to the height of three or foure cubits, especially when it is set by a wall : it consisteth of slender brittle branches, whereon do grow very many long leaves, narrow, somewhat hard, of a quicke spicy taste, whitish underneath, and of a full greene colour above, or in the upper side, with a pleasant sweet strong smell; among which come forth little floures of a whitish blew colour: the seed is blackish: the roots are tough and wooddy.

Rosemary groweth in France, Spaine, and in other hot countries; in woods, and in untilled places: there is such plenty thereof in Languedocke, that the inhabitants burne scarce any other fuell: they make hedges of it in the gardens of Italy and England, being a great ornament unto the same: it groweth neither in the fields nor gardens of the Easterne cold countries; but is carefully and curiously kept in pots, set into the stoves and cellers, against the injuries of their cold Winters.

Rosemary floureth twice a yeare, in the Spring, and after in August. It is called in Latine, *Rosemarinus Coronaria*: it is surnamed *Coronaria*, because women have beene accustomed to make crownes and garlands thereof.

The distilled water of the floures of Rosemary being drunke at morning and evening first and last, taketh away the stench of the mouth and breath, and maketh it very sweet, if there be added thereto, to steep or infuse for certaine daies, a few Cloves, Mace, Cinnamon, and a little Annise seed.

The Arabians and other Physitions succeeding, do write, that Rosemary comforteth the braine, the memorie, the inward senses, and restoreth speech unto them that are possessed with the dumbe palsie, especially the conserve made of the floures and sugar, or any other way confected with sugar, being taken every day fasting.

The floures made up into plates with Sugar after the manner of Sugar Roset and eaten, comfort the heart, and make it merry, quicken the spirits, and make them more lively.

Worts or Wortle berries

Vaccinia nigra, the blacke Wortle or Hurtle, is a base and low shrub or wooddy plant, bringing forth many branches of a cubit high, set full of small leaves of a darke greene colour, not much unlike the leaves of Box or the Myrtle tree: amongst which come forth little hollow floures turning into small berries, greene at the first, afterward red, and at the last of a blacke colour, and full of a pleasant and sweet juyce: in which doe lie divers little thinne whitish seeds: these berries do colour the mouth and lips of those that eat them, with a blacke colour: the root is wooddy, slender, and now and then creeping.

Vaccinia rubra, or red Wortle, is like the former in the manner of growing, but that the leaves are greater and harder, almost like the leaves of the Box tree, abiding greene all the Winter long: among which come forth small carnation floures, long and round, growing in clusters at the top of the branches: after which succeed small berries, in shew and bignesse like the former, but that they are of an excellent red colour and full of juyce, of so orient and beautifull a purple to limme withall, that Indian *Lacca* is not to be compared thereunto, especially when this juyce is prepared and dressed with Allom according to art, as my selfe have proved by experience: the taste is rough and astringent: the root is of a wooddy substance.

These plants prosper best in a leane barren soile, and in untoiled wooddy places: they are now and then found on high hills subject to the winde, and upon mountaines: they grow plentifully in both the Germanies, Bohemia, and in divers places of France and England; namely in Middlesex on Hampsted heath, and in the woods thereto adjoyning, and also upon the hills in Cheshire called Broxen hills, neere Beeston castle, seven miles from the

Nantwich; and in the wood by Highgate called Finchley wood, and in divers other places.

The Wortle berries do floure in May, and their fruit is ripe in June.

The people of Cheshire do eat the blacke Wortles in creame and milke, as in these South parts we eate Strawberries.

GOOSE-BERRIE, OR FEA-BERRY BUSH

Goose-Berry

There be divers sorts of the Goose-berries; some greater, others lesse: some round, others long; and some of a red colour: the figure of one shall serve for the rest.

The Goose-berry bush is a shrub of three or foure cubits high, set thicke with most sharpe prickles: it is likewise full of branches, slender, wooddy, and prickly: whereon do grow round leaves cut with deepe gashes into divers parts like those of the Vine, of a very greene colour: the floures be very small, of a whitish greene, with some little purple dashed here and there: the fruit is round, growing scatteringly upon the branches, greene at the first, but waxing a little yellow through maturitie; full of a winie juyce somewhat sweet in taste when they be ripe; in which is contained hard seed of a whitish colour: the root is wooddy, and not without strings annexed thereto.

There is another whose fruit is almost as big as a small Cherry, and very round in forme: as also another of the like bignesse, of an inch in length, in taste and substance agreeing with the common sort.

We have also in our London gardens another sort altogether without prickles: whose fruit is very smal, lesser by much than the common kinde, but of a perfect red colour, wherein it differeth from the rest of his kinde.

These plants doe grow in our London Gardens and else-where in great abundance. The leaves come forth in the beginning of Aprill or sooner: the fruit is ripe in June and July.

This shrub hath no name among the old Writers, who as we deeme knew it not, or else esteemed it not: in English, Goose-berry, Goose-berry bush, and Fea-berry bush in Cheshire, my native country.

The fruit is used in divers sauces for meat, as those that are skilfull in cookerie can better tell than my selfe. They are used in broths in stead of Verjuice, which maketh the broth not onely pleasant to the taste, but is greatly profitable to such as are troubled with an hot burning ague.

They are diversly eaten, but they every way ingender raw and cold bloud: they nourish nothing or very little.

STRAW-BERRIES

There be divers sorts of Straw-berries; one red, another white, a third sort greene, and likewise a wilde Strawberry, which is altogether barren of fruit.

Straw-berries do grow upon hills and vallies, likewise in woods and other such places that bee somewhat shadowie: they prosper well in Gardens.

The fruit or berries are called in Latine by *Virgil* and *Ovid*, *Fraga*: neither have they any other name commonly knowne: in French, *Fraises*: in English, Straw-berries.

The leaves boyled and applied in manner of a pultis taketh away the burning heate in wounds: the decoction thereof strengthneth the gummes, and fastneth the teeth.

The distilled water drunke with white Wine is good against the passion of the heart, reviving the spirits, and making the heart merry.

The ripe Strawberries quench thirst, and take away, if they be often used, the rednesse and heate of the face.

MEDOW-GRASSE

There be sundry and infinite kindes of Grasses not mentioned by the Antients, either as unnecessarie to be set downe, or unknowne to them: only they make mention of some few, whose wants we meane to supply, in such as have come to our knowledge, referring the rest to the curious searcher of Simples.

Straw-Berries

Common Medow Grasse hath very small tufts or roots, with thicke hairy threds depending upon the highest turfe, matting and creeping on the ground with a most thicke and apparant shew of wheaten leaves, lifting up long thinne jointed and light stalks, a foot or a cubit high, growing small and sharpe at the top, with a loose ear hanging downward, like the tuft or top of the common Reed.

Small Medow Grasse differeth from the former in the

139

varietie of the soile; for as the first kinde groweth in medowes, so doth this small Grasse clothe the hilly and more dry grounds untilled, and barren by nature; a Grasse more fit for sheepe than for greater cattell. And because the kindes of Grasse do differ apparantly in root, tuft, stalke, leafe, sheath, eare, or crest, we may assure our selves that they are endowed with severall Vertues,

formed by the Creator for the use of man, although they have been by a common negligence hidden and unknowne. And therefore in this our Labor we have placed each of them in their severall bed, where the diligent searcher of Nature, may if so he please, place his learned observations.

Common Medow-Grasse groweth of it selfe unset or unsowne, every where; but the small Medow-Grasse for the most part groweth upon dry and barren grounds, as partly we have touched in the Description.

Concerning the time when Grasse springeth and seed-

Medow-Grasse

eth, I suppose there is none so simple but knoweth it, and that it continueth all the whole yeare, seeding in June and July. Neither needeth it any propagation or replanting by seed or otherwise; no not so much as the watery Grasses, but that they recover themselves againe, although they have beene drowned in water all the Winter long, as may appeare in the wilde fennes in Lincolnshire, and such like places.

Grasse groweth, goeth, or spreadeth it selfe unset or unsowne naturally over all fields or grounds, cloathing them with a faire and perfect green. It is yearely mowed, in some places twice, and in some rare places thrice. Then is it dried and withered by the heate of the Sun, with often turning it; and then is it called in English, Hay: in French, *Le herbe du praiz.*

R E E D S

Of Reeds the Ancients have set downe many sorts. The common Reed hath long strawie stalkes, full of knotty joints or knees like unto corne, whereupon doe grow very long rough flaggy leaves. The tuft or spokie eare doth grow at the top of the stalkes, browne of colour, barren and without seed, and doth resemble a bush of feathers, which turneth into fine downe or cotton which is caried away with the winde. The root is thicke, long, and full of strings, dispersing themselves farre abroad, whereby it doth greatly increase.

The great sort of Reeds or Canes hath no particular description to answer your expectation, for that as yet there is not any man which hath written thereof, especially of the manner of growing of them, either of his owne knowledge or report from others, so that it shall suffice that he know that that great cane is used especially in Constantinople and thereabout, of aged and wealthy Citisens, and also Noblemen and such great personages, to make them walking staves of, carving them at the top with sundry Scutchions, and pretty toyes of imagerie for the beautifying of them; and so they of the better sort doe garnish them both with silver and gold.

The common Reed groweth in standing waters and in the edges and borders of rivers almost every where; and the other being the angling Cane for fishers groweth in Spaine and those hot Regions.

They flourish and floure from Aprill to the end of September, at which time they are cut downe for the use of man, as all do know.

The roots of reed stamped small draw forth thorns and splinters fixed in any part of mans body. The same stamped with vinegre ease all luxations and members out of joynt. And likewise stamped they heale hot and sharpe inflammations. The ashes of them mixed with vinegre helpe the scales and scurfe of the head, and the falling of the haire.

The great Reed or Cane is not used in physicke, but is esteemed to make slears for Weavers, sundry sorts of pipes, as also to light candles that stand before Images, and to make hedges and pales, as we do of lats and such like; and also to make certain divisions in ships to divide the sweet oranges from the sowre, the Pomecitron and lemmons likewise in sunder, and many other purposes.

Paper Reed

Paper Reed hath many large flaggie leaves somewhat triangular and smooth, not much unlike those of Catstaile, rising immediately from a tuft of roots compact of many strings, amongst the which it shooteth up two or three naked stalkes, square, and rising some six or seven cubits high above the water: at the top whereof there stands a tuft or bundle of chaffie threds set in comely order, resembling a tuft of floures, but barren and void of seed.

This kinde of Reed growes in the Rivers about Babylon, and neere the city Alcaire, in the river Nilus, and such other places of those countries.

This Reed, which I have Englished Paper Reed, or Paper plant, is the same (as I doe reade) that Paper was made of in Ægypt, before the invention of paper made of linnen clouts was found out. It is thought by men of great learning and understanding in the Scriptures, and set downe by them for truth, that this plant is the same Reed mentioned in the second chapter of *Exodus*; whereof was made that basket or cradle, which was dawbed within and without with slime of that countrey, called *Bitumen Judaicum*, wherein *Moses* was put being committed to the water, when *Pharaoh* gave commandement that all the male children of the Hebrewes should be drowned.

Paper Reed

The roots of Paper Reed doe nourish, as may appeare by the people of Ægypt, which doe use to chew them in their mouthes, and swallow downe the juice, finding therein great delight and comfort.

BURRE-REED

The first of these plants hath long leaves, which are double edged, or sharpe on both sides, with a sharpe crest in the middle, in such manner raised up that it seemeth to be triangle or three-square. The stalks grow among the leaves, and are two or three foot long, being divided into many branches, garnished with many prickly husks or knops of the bignesse of a nut. The root is full of hairy strings.

The great water Burre differeth not in any thing from the first kinde in roots or leaves, save that the first hath his leaves rising immediatly from the tuft or knop of the root; but this kinde hath a long stalke comming from the root, whereupon a little above the root the leaves shoot out round about the stalke successively, some leaves still growing above others, even to the top of the stalke, and from the top thereof downward by certaine distances. It is garnished with many round wharles or rough coronets, having here and there among the said wharles one single short leafe of a pale greene colour.

Branched Burre Reed

Both these are very common, and grow in moist medowes and neere unto water courses. They plentifully grow in the fenny grounds of Lincolnshire and such like places; in the ditches about S. *Georges* fields, and in the ditch right against the place of execution at the end of Southwark, called S. *Thomas* Waterings.

Some call the first *Sparganium ramosum*, or branched Burre Reed.

Jasmine, or Gelsemine

Jasmine, or Gelsemine, is of the number of those plants which have need to be supported or propped up, and yet notwithstanding of it selfe claspeth not or windeth his stalkes about such things as stand neere unto it, but onely leaneth and lieth upon those things that are prepared to sustain it about arbors and banqueting houses in gardens, by which it is held up: the stalkes thereof are long, round, branched, jointed or kneed, and of a green colour, having within a white spongeous pith. The leaves stand upon a middle rib, set together by couples like those of the ash tree, but much smaller, of a deepe greene colour: the floures grow at the uppermost part of the branches, standing in a small tuft far set one from another, sweet in smell, of colour white: the seed is flat and broad like those of Lupines, which seldom come to ripenesse: the root is tough and threddy.

The oile which is made of the flours hereof wasteth away raw humors, and is good against cold rheums; but in those that are of a hot constitution it causeth head-ache, and the overmuch smell thereof maketh the nose to bleed. It is good to be anointed after baths, in those bodies that have a need to be suppled and warmed.

Lavander Spike

Lavander Spike hath many stiffe branches of a wooddy substance, growing up in the manner of a shrub, set with many long hoarie leaves, by couples for the most part, of a strong smell, and yet pleasant enough to such as do love strong savors. The floures grow at the top of the branches, spike fashion, of a blew colour.

The second differeth not from the precedent; but in the colour of the floures: For this plant bringeth milke white floures; and the other blew, wherein especially consisteth the difference.

Lavander

We have in our English gardens a small kinde, which is altogether lesser than the other. Lavander Spike is called in Latine *Lavendula*, and *Spica*: in Spanish, *Spigo*, and *Languda*. The first is the male, and the second the female. It is thought of some to bee that sweet herbe *Casia*, whereof *Virgil* maketh mention in the second Eclog of his Bucolicks:

And then shee'l Spike and
    such sweet hearbs infold
And paint the Jacinth with
    the Marigold.

And likewise in the fourth of his Georgickes, where he intreateth of chusing of seats and places for Bees, and for the ordering thereof, he saith thus:

About them let fresh Lavander and store
Of wilde Time with strong Savorie to floure.

The distilled water of Lavander smelt unto, or the temples and forehead bathed therewith, is a refreshing to them that have the Catalepsy, a light migram, and to them that have the falling sicknesse, and that use to swoune much. But when there is abundance of humours, it is not then to be used safely, neither is the composition to be taken which is made of distilled wine: in which such kinds of herbes, floures, or seeds, and certain spices are infused or steeped, though most men do rashly and at adventure give them without making any difference at al. For by using such hot things that fill and stuffe the head,

146

both the disease is made greater, and the sick man also brought into daunger, especially when letting of bloud, or purging have not gon before. Thus much by way of admonition, because that every where some unlearned Physitians and divers rash & overbold Apothecaries, and other foolish women, do by and by give such compositions, and others of the like kind, not only to those that have the Apoplexy; but also to those that are taken, or have the Catuche or Catalepsis with a Fever; to whom they can give nothing worse, seeing those things do very much hurt, and oftentimes bring death it selfe.

The floures of Lavander picked from the knaps, I meane the blew part and not the husk, mixed with Cinnamon, Nutmegs, & Cloves, made into pouder, and given to drinke in the distilled water thereof, doth helpe the panting and passion of the heart, prevaileth against giddinesse, turning, or swimming of the braine, and members subject to the palsie.

Conserve made of the floures with sugar, profiteth much against the diseases aforesaid, if the quantitie of a beane be taken thereof in the morning fasting.

CLOVE GILLOFLOURES

There are at this day under the name of *Cariophyllus* comprehended divers and sundry sorts of plants, of such various colours, and also severall shapes, that a great and large volume would not suffice to write of every one at large in particular; considering how infinite they are, and how every yeare every clymate and country bringeth forth new sorts, such as have not heretofore been written of; some whereof are called Carnations, others Clove Gillofloures, some Sops in wine, some Pagiants, or Pagion color, Horse-flesh, blunket, purple, white, double and single Gillofloures, as also a Gillofloure with yellow flours: the which a worshipful Merchant of London

Mr. *Nicholas Lete* procured from Poland, and gave me thereof for my garden, which before that time was never seen nor heard of in these countries.

The great Carnation Gillo-floure hath a thick round wooddy root, from which riseth up many strong joynted stalks set with long green leaves by couples: on the top of the stalks do grow very fair floures of an excellent sweet smell, and pleasant Carnation colour, whereof it tooke his name.

The Clove Gillofloure differeth not from the Carnation but in greatnesse as well of the flowres as leaves. The floure is exceeding well knowne, as also the Pinkes and other Gillofloures; wherefore I will not stand long upon the description.

These Gillofloures, especially the Carnations, are kept in pots from the extremitie of our cold Winters. The Clove Gillofloure endureth better the cold, and therefore is planted in gardens.

The Clove Gillofloure is called of the later Herbarists *Caryophylleus Flos*, of the smell of cloves wherewith it is possessed.

The double Clove Gillofloure

Johannes Ruellius saith, That the Gillofloure was unknowne to the old writers: whose judgement is very good, especially because this herb is not like to that of *Vetonica*, or *Cantabrica*. It is marvell, saith he, that such a famous floure, so pleasant & sweet, should lie hid, and not be made known by the old writers: which may be thought not inferior to the rose in beautie, smell, and varietie.

The conserve made of the floures of the Clove Gillofloure and sugar, is exceeding cordiall, and wonderfully above measure doth comfort the heart, being eaten now and then.

MEDE-SWEET, OR QUEENE OF THE MEDOWES

This herbe hath leaves like Agrimony, consisting of divers leaves set upon a middle rib like those of the ash tree, every small leaf sleightly snipt about the edges, white on the inner side, and on the upper side crumpled or wrinkled like unto those of the Elme tree; whereof it tooke the name *Ulmaria*, of the similitude or likenesse that the leaves have with the Elme leaves. The stalke is three or foure foot high, rough, and very fragile or easie to bee broken, of a reddish purple colour: on the top whereof are very many little floures clustering and growing together, of a white colour tending to yellownesse, and

Queene of the Medow

of a pleasant sweet smell, as are the leaves likewise.

It groweth in the brinkes of waterie ditches and rivers sides, and also in medowes: it liketh watery and moist places, and groweth almost every where. It floureth and flouresheth in June, July, and August.

It is called of the later age *Regina prati*: in English, Meads-sweet, Medow-sweet, and Queen of the medowes.

It is reported, that the floures boiled in wine and drunke, do make the heart merrie.

The leaves and floures farre excell all other strowing herbes, for to decke up houses, to straw in chambers, halls, and banqueting houses in the Summer time; for the smell thereof makes the heart merrie, delighteth the senses: neither doth it cause headache, or lothsomenesse to meat, as some other sweet smelling herbes do.

The distilled water of the floures dropped into the eies, taketh away the burning and itching thereof, and cleareth the sight.

WATER-FERNE, OR OSMUND THE WATER-MAN

Water-Ferne hath a great triangle stalke two cubits high, beset upon each side with large leaves spred abroad like wings, and dented or cut like Polypody: these leaves are like the large leaves of the Ash-tree; for doubtlesse when I first saw them a far off it caused me to wonder thereat, thinking that I had seene young Ashes growing upon a bog, but beholding it a little neerer, I might easily distinguish it from the Ash, by the browne rough and round graines that grew on the top of the branches, which yet are not the seed thereof, but are very like unto the seed. The root is great and thicke, folded and covered over with many scales and interlacing roots, having in the middle of the great and hard wooddy part thereof some small whitenesse, which hath beene called the heart of *Osmund* the water-man.

It groweth in the midst of a bog at the further end of Hampsted heath from London, at the bottome of a hill adjoyning to a small cottage, and in divers other places, as also upon divers bogges on a heath or a common neere unto Bruntwood in Essex, especially neere unto a place there that some have digged, to the end to finde a nest or mine of gold; but the birds were over fledge, and flowne away before their wings could be clipped.

It is called in English, Water-Ferne, Osmund the Water-man: of some, Saint Christophers herbe, and Osmund.

The root and especially the heart or middle part thereof boyled or else stamped, and taken with some kinde of liquor, is thought to be good for those that are wounded, dry-beaten and bruised; that have fallen from some high place: and for the same cause the Empericks do put it in decoctions, which the later Physitians doe call wound-drinkes: some take it to be so effectuall, and of so great a vertue, as that it can dissolve cluttered bloud remaining in any inward part of the body, and that it also can expell or drive it out by the wound.

The tender sprigs thereof at their first comming forth are excellent good unto the purposes aforesaid, and are good to be put into balmes, oyles, and consolidatives, or healing plaisters, and into unguents appropriate unto wounds, punctures, and such like.

SPLEENE-WOORT OR MILT-WASTE

Spleen-wort, being that kinde of Fern called *Asplenium* or *Ceterach*, and the true *Scolopendria*, hath leaves a span long, jagged or cut upon both sides, even hard to the middle rib, every cut or incisure being as it were cut halfe round (whereby it is knowne from the rough Spleenwort) not one cut right against another, but one besides the other, set in several order, being slipperie and green on the upper side, soft and downy underneath; which when they bee withered are folded up together like a scrole, and hairy without, much like to the rough Bear-worme wherewith men bait their hooks to catch fish. The root is small, blacke, and rough, much platted or interlaced, having neither stalke, floure, nor seeds.

Rough Spleenwort is partly like the other Ferns in shew, and beareth neither stalk nor seed, having narrow leaves a foot long and somewhat longer, slashed on the edges even to the middle rib, smooth on the upper side, and of a swart green colour underneath.

Rough Spleenwort

Ceterach groweth upon old stone walls and rocks in darke and shadowie places throughout the West parts of England.

The rough Spleenwort groweth upon barren heaths, dry sandy bankes, and shadowie places in most parts of England, but especially on a heath by London called Hampsted heath, where it grows in great aboundance.

There be Empericks or blinde practitioners of this age who teach, that with this herb not onely the hardnesse and swelling of the spleene, but all infirmities of the liver also may be effectually, and in very short time removed, insomuch that the sodden liver of a beast is restored to his former constitution againe, that is, made like unto a raw liver, if it be boiled again with this herb.

But this is to be reckoned among the old Wives fables, and that also which *Dioscorides* tells of, touching the gathering of Spleene-wort in the night, and other most vain things, which are found here and there scattered in old books: from which most of the later Writers do not abstaine, who many times fill up their pages with lies and frivolous toyes, and by so doing do not a little deceive yong Students.

M o o n e - w o r t

The small Lunary springeth forth of the ground with one leafe like Adders-tongue, jagged or cut on both sides into five or six deepe cuts or notches, not much unlike the leaves of *Scolopendria*, or *Ceterach*, of a greene colour; whereupon doth grow a small naked stem of a finger long, bearing at the top many little seeds clustering together; which being gathered and laid in a platter or such like thing for the space of three weekes, there will fall from the same a fine dust or meale of a whitish colour, which is the seed if it bring forth any. The root is slender, and compact of many small threddy strings.

Small Moone-woort is singular to heale greene and fresh wounds. It hath beene used among the Alchymists and witches to doe wonders withall, who say, that it will loose lockes, and make them to fall from the feet of horses that grase where it doth grow, and hath beene called of them *Martagon*, whereas in truth they are all but drowsie dreames and illusions; but it is singular for wounds as aforesaid.

S c o r p i o n   G r a s s e

Scorpion grasse hath many smooth, plaine, even leaves, of a darke greene colour; stalkes small, feeble and weake, trailing upon the ground, and occupying a great circuit in respect of the plant. The floures grow upon long and slender foot-stalks, of colour yellow, in shape like to the floures of broome; after which succeed long, crooked, rough cods, in shape and colour like unto a Caterpiller; wherein is contained yellowish seed like unto a kidney in shape. The root is small and tender: the whole plant perisheth when the seed is ripe.

There is another sort almost in every shallow gravelly running streame, having floures blew of colour and sometimes with a spot of yellow among the blew.

There is likewise another sort growing upon most dry gravelly and barren ditch bankes, with leaves like those of Mouse-eare: this is called *Myosotes scorpioides*: it hath rough and hairy leaves, of an overworne russet colour: the floures doe grow upon weake, feeble, and rough branches, as is all the rest of the plant. They grow for the most part at one side of the stalke, blew of colour, with a like little spot of yellow as the other, turning themselves backe againe like the tail of a Scorpion.

These Scorpion grasses grow not wilde in England, notwithstanding I have received seed of the first from beyond the seas, and have dispersed them through England,

Mouse-eare Scorpion grasse

which are esteemed of gentlewomen for the beautie and strangnesse of the crooked cods resembling Caterpillers.

Dioscorides saith, that the leaves of Scorpion grasse applyed to the place, are a present remedy against the stinging of Scorpions: and likewise boyled in wine and drunke, prevaile against the said bitings, as also of addars, snakes, and such venomous beasts: being made in an unguent with oile, wax, and a little gum *Elemni*, they are profitable against such hurts as require an healing medicine.

SLEEPY NIGHTSHADE

Dwale or sleeping Nightshade hath round blackish stalkes six foot high, whereupon do grow great broad leaves of a dark green colour: among which grow smal hollow floures bel-fashion, of an overworn purple colour; in the place whereof come forth great round berries of the bignesse of the black chery, green at the first, but when they be ripe of the colour of black jet or burnished horne, soft, and ful of purple juice; among which juice lie the seeds, like the berries of Ivy: the root is very great, thick, and long lasting.

This kinde of Nightshade causeth sleep, troubleth the mind, bringeth madnesse if a few of the berries be inwardly taken, but if moe be given they also kill and bring present death. If you will follow my counsell, deale not with the same in any case, and banish it from your gardens and the use of it also, being a plant so furious and deadly: for it bringeth such as have eaten thereof into a dead sleepe wherein many have died, as hath beene often seene and proved by experience both in England and elsewhere. But to give you an example hereof it shall not be amisse: It came to passe that three boies of Wisbich in the Isle of Ely did eate of the pleasant and beautifull fruit hereof, two whereof died in lesse than eight houres after that they had eaten of them. The third child had a quantitie of honey and water mixed together given him to drinke, causing him to vomit often: God blessed this meanes and the child recovered.

COTTON-WEED OR CUD-WEED

English Cudweed hath sundry slender and upright stalks divided into many branches, and groweth as high as common Wormwood, whose colour and shape it much resembleth. The leaves shoot from the bottome of the

155

turfe full of haires, among which do grow small pale
coloured floures like those of the smal *Coniza* or Flea-
bane. The whole plant is of a bitter taste.

There is a kinde of Cotton-weed, being of greater
beautie than the rest, that hath strait and upright stalkes 3
foot high or more, covered with a most soft and fine wooll,
and in such plentifull manner, that a man may with his
hands take it from the stalke in great quantitie: which

Herbe impious, or wicked Cudweed

stalke is beset with many
small long and narrow leaves,
greene upon the inner side,
and hoary on the other side,
fashioned somewhat like the
leaves of Rosemary, but
greater. The floures do grow
at the top of the stalkes in
bundles or tufts, consisting of
many small floures of a white
colour, and very double, compact, or as it were consisting
of little silver scales thrust
close together, which doe
make the same very double.
When the floure hath long
flourished, and is waxen old,
then comes there in the
middest of the floure a certaine
browne yellow thrumme, such
as is in the middest of the
Daisie: which floure being
gathered when it is young, may be kept in such manner as
it was gathered (I meane in such freshnesse and well
liking) by the space of a whole yeare after, in your chest or
elsewhere: wherefore our English women have called it
Live-long, or Live for ever, which name doth aptly
answer his effects.

Small Cudweed hath three or foure small grayish cottony or woolly stalkes, growing strait from the root, and commonly divided into many little branches: the leaves be long, narrow whitish, soft and woolly, like the other of his kinde: the floures be round like buttons, growing very many together at the top of the stalkes, but nothing so yellow as Mouse-eare, which turne into downe, and are caried away with the winde.

Wicked Cudweed is like unto the last before mentioned, in stalkes, leaves, and floures, but much larger, and for the most part those floures which appeare first are the lowest, and basest, and they are overtopt by other floures which come on younger branches, and grow higher, as children seeking to overgrow or overtop their parents, (as many wicked children do) for which cause it hath beene called *Herba impia*, that is, the wicked Herbe, or Herbe Impious.

The fume or smoke of the herbe dried, and taken with a funnell, being burned therein, and received in such manner as we use to take the fume of Tabaco, that is, with a crooked pipe made for the same purpose by the Potter, prevaileth against the cough of the lungs, the great ache or paine of the head, and cleanseth the breast and inward parts.

FETHERFEW

Feverfew bringeth forth many little round stalkes, divided into certaine branches. The leaves are tender, diversly torne and jagged, and nickt on the edges like the first and nethermost leaves of Coriander, but greater. The floures stand on the tops of the branches, with a small pale of white leaves, set round about a yellow ball or button, like the wild field Daisie. The root is hard and tough: the whole plant is of a light whitish greene colour, of a strong smell, and bitter taste.

The common single Feverfew groweth in hedges, gardens, and about old wals, it joyeth to grow among rubbish. There is oftentimes found when it is digged up a little cole under the strings of the root, and never without it, whereof *Cardane* in his booke of Subtilties setteth down divers vaine and trifling things.

Feverfew dried and made into pouder, and two drams of it taken with hony or sweet wine, purgeth by siege melancholy and flegme; wherefore it is very good for them that are giddie in the head, or which have the turning called *Vertigo*, that is, a swimming and turning in the head. Also it is good for such as be melancholike, sad, pensive, and without speech.

M ULLEIN

The male Mullein or Higtaper hath broad leaves, very soft, whitish and downy; in the midst of which riseth up a stalk, straight, single, and the same also whitish all over, with a hoary down, and covered with the like leaves, but lesser and lesser even to the top; among which taperwise are set a multitude of yellow floures consisting of five leaves apiece: in the places wherof come up little round vessels, in which is contained very small seed. The root is long, a finger thicke, blacke without, and full of strings.

The female Mullein hath likewise many white woolly leaves, set upon an hoary cottony upright stalke of the height of foure or five cubits: the top of the stalke resembleth a torch decked with infinite white floures, which is the speciall marke to know it from the male kinde, being like in every other respect.

These plants grow of themselves neere the borders of pastures, plowed fields, or causies & dry sandy ditch banks, and in other untilled places. They grow in great plenty neere unto a lyme-kiln upon the end of Blacke heath next to London, as also about the Queenes house at

Eltham neere to Dartford in Kent; in the highwayes about
Highgate neere London, and in most countries of England
that are of a sandy soile.

They are found with their floure from July to
September, and bring forth their
seed the second yeare after it is
sowne.

Mullein is called in English
Mullein, or rather Woollen,
Higtaper, Torches, Longwort,
and Bullocks Longwort; and of
some, Hares beard.

The country people, especially
the husbandmen in Kent, do give
their cattel the leaves to drink
against the cough of the lungs,
being an excellent approved
medicine for the same, wherupon they call it Bullocks
Lungwort.

The report goeth (saith *Pliny*)
that figs do not putrifie at all
that are wrapped in the leaves of
Mullein.

White floured Mullein

GOATS BEARD, OR GO TO BED AT NOONE

Goats-beard, or Go to bed at noone hath hollow stalks,
smooth, and of a whitish green colour, whereupon do
grow long leaves crested downe the middle with a
swelling rib, sharp pointed, yeelding a milkie juice when
it is broken, in shape like those of Garlick: from the
bosome of which leaves thrust forth smal tender stalks, set
with the like leaves, but lesser: the floures grow at the
top of the stalks, consisting of a number of purple leaves,
dasht over as it were with a little yellow dust, set about

with nine or ten sharp pointed green leaves: the whole floure resembles a Star when it is spred abroad; for it shutteth it selfe at twelve of the clock, and sheweth not his face open untill the next daies Sunne doth make it floure anew, whereupon it was called Go to bed at noone: when these floures be come to their full maturitie and ripenesse, they grow into a downy Blow-ball like those of Dandelion, which is carried away with the winde.

The yellow Goats beard hath the like leaves, stalks, root, seed, and downie blow-balls that the other hath, and also yeeldeth the like quantitie of milke, insomuch that if the pilling while it is greene be pulled from the stalks, the milky juice followeth: but when it hath there remained a little while it waxeth yellow. The floures hereof are of a gold yellow colour.

The first growes not wild in England that I could ever see or heare of, except in Lancashire on the banks of the river Chalder, neere to my Lady *Heskiths* house, two miles from Whawley: it is sown in gardens for the beauty of the

Goats-Beard

floures almost every where. The other growes plentifully in most of the fields about London, and in divers other places.

Goats-beard is called in English, Joseph's floure, Star of Jerusalem, Noon tide, and Go to bed at noone.

The roots of Goats-beard boiled in wine and drunk, asswageth the pain and pricking stitches of the sides.

The same boiled in water untill they be tender, and buttered as parnseps and carrots, are a most pleasant and wholesome meate, in delicate taste far surpassing either Parsenep or Carrot: which meat procures appetite, warmeth the stomacke, and strengthneth those that have been sicke of a long lingring disease.

A N G E L I C A

Angelica is very common in our English gardens; in other places it growes wild without planting, as in Norway, and in an Island of the North called Island, where it groweth very high; it is eaten of the inhabitants, the bark being pilled off, as we understand by some that have travelled into Island, who were sometimes compelled to eat hereof for want of other food; and they report that it hath a good and pleasant taste to them that are hungry. It groweth likewise in divers mountains of Germanie, and especially of Bohemia.

The root of garden Angelica is a singular remedy against poyson, and against the plague, and all infections taken by evill and corrupt aire; if you doe but take a piece of the root and hold it in your mouth, or chew the same between your teeth, it doth most certainely drive away the pestilentiall aire, yea although the corrupt aire have possessed the hart, yet it driveth it out againe, as Rue and Treacle do.

W I L D   C L A R I E ,   O R   O C U L U S   C H R I S T I

Oculus Christi is a kinde of Clarie, but lesser: the stalkes are many, a cubit high, squared and somewhat hairie; the leaves be broad, rough, and of a blackish greene colour. The floures grow alongst the stalkes, of a blewish colour. The seed is round and blackish, the root is thicke and tough, with some threds annexed thereto.

The purple Clarie hath leaves somewhat round, layed over with a hoary cottony substance, not much unlike Horehound: among which rise up small hairy square stalkes, set toward the top with little leaves of a purple colour, which appeare at the first view to be flours, and yet are nothing else but leaves, turned into an excellent purple colour: and among these beautifull leaves come

Purple Clarie

forth smal floures of a blewish or watched colour, in fashion like to the floures of Rosemarie; which being withered, the husks wherein they do grow containe certaine blacke seed, that falleth forth upon the ground very quickly, because that every such huske doth turne and hang downe his head toward the ground. The root dieth at the first approch of Winter.

The first groweth wilde in divers barren places, almost in every country, especially in the fields of Holborne neere unto Grayes Inne, in the high way by the end of a brickewall: at the end of Chelsey next to London, in the high way as you go from the Queenes pallace of Richmond to the waters side, and in divers other places. The other is a stranger in England: it groweth in my Garden.

Wilde Clarie is called after the Latine name *Oculus Christi*, of his effect in helping the diseases of the eies. The seed put whole into the eies, clenseth and purgeth them exceedingly from waterish humors, rednesse, inflammation, and divers other maladies, or all that happen unto

162

the eies, and takes away the paine and smarting thereof, especially being put into the eies one seed at one time, and no more, which is a generall medicine in Cheshire and other countries thereabout, knowne of all, and used with good success.

S A G E

We have in our gardens a kinde of Sage, the leaves whereof are reddish; part of those red leaves are stripped with white, others mixed with white, greene, and red, even as Nature list to play with such plants. This is an elegant variety, and is called *Salvia variegata elegans*, Variegated or painted Sage.

We have also another, the leaves whereof are for the most part white, somewhat mixed with greene, often one leafe white, and another greene, even as Nature list, as we have said. This is not so rare as the former, nor neere so beautifull, wherefore it may be termed *Salvia variegata vulgaris*, Common painted Sage.

Sage is singular good for the head and braine; it quickneth the sences and memory, strengthneth the sinewes, restoreth health to those that have the palsie, takes away shaking or trembling of the members; and being put up into the nosthrils, it draweth thin flegme out of the head.

It is likewise commended against the spitting of bloud, the cough, and paines of the sides, and bitings of Serpents.

No man needs to doubt of the wholesomnesse of Sage Ale, being brewed as it should be, with Sage, Scabious, Betony, Spikenard, Squinanth, and Fennell seeds.

The leaves of red Sage put into a woodden dish, wherein is put very quicke coles, with some ashes in the bottome of the dish to keepe the same from burning, and a little vinegre sprinkled upon the leaves lying upon the coles, and so wrapped in linnen cloath, and holden very hot unto the side of those that are troubled with a grievous stitch, taketh away the paine presently.

BAWME

Apiastrum, or *Melissa*, is our common best knowne Balme or Bawme, having many square stalkes and blackish leaves, of a pleasant smell, drawing neere in smell and savour unto a Citron: the floures are of a Carnation colour.

Bawme is much sowen and set in Gardens, and oftentimes it groweth of it selfe in Woods and mountaines, and other wilde places: it is profitably planted in Gardens, as *Pliny* writeth, about places where Bees are kept, because they are delighted with this herbe above others, whereupon it hath beene called *Apiastrum*: for, saith he, when they are straied away, they

Bastard Bawme with white floures

doe finde their way home againe by it, as *Virgil* writeth in his Georgicks:

———Here liquors cast in fitting sort,
Of bruised Bawme and more base Honywort.

Bawme drunke in wine is good against the bitings of venomous beasts, comforts the heart, and driveth away all melancholy and sadnesse.

The hives of Bees being rubbed with the leaves of Bawme, causeth the Bees to keep together, and causeth others to come unto them.

The later age, together with the Arabians and Mauritanians, affirme Balme to be singular good for the heart and to be a remedy against the infirmities thereof; for *Avicen* in his booke written of the infirmities of the heart, teacheth that Bawme makes the heart merry and joyfull, and strengtheneth the vitall spirits.

Dioscorides writeth, That the leaves drunke with wine, or applied outwardly, are good against the stingings of venomous beasts, and the biting of mad dogs; also it helpeth the tooth-ache, the mouth being washed with the decoction, and is likewise good for those that cannot take breath unlesse they hold their neckes upright.

Smiths Bawme or Carpenters Bawme is most singular to heale up greene wounds that are cut with yron; it cureth the rupture in short time. *Pliny* saith that it is of so great vertue, that though it be but tied to his sword that hath given the wound, it stancheth the bloud.

BORAGE

Borage hath broad leaves, rough, lying flat upon the ground, of a blacke or swart green colour: among which riseth up a stalke two cubits high, divided into divers branches, wherupon do grow gallant blew floures, composed of five leaves apiece; out of the middle of which grow forth blacke threds joined in the top, and pointed like a broch or pyramide: the root is threddy, and cannot away with the cold of Winter.

Borage with white floures is like unto the precedent, but differeth in the floures, for those of this plant are white, and other of a perfect blew colour, wherein is the difference.

These grow in my garden and in others also. Borage

floures and flourishes most part of all Summer, and till Autumne be far spent.

Borage is called in shops *Borago*: *Pliny* calleth it *Euphrosinum*, because it makes a man merry and joyfull: which thing also the old verse concerning Borage doth testifie:

> *Ego Borago gaudia semper ago.*
> I Borage bring alwaies courage.

Those of our time do use the floures in sallads, to exhilerate and make the minde glad. There be also many things made of them, used for the comfort of the heart, to drive away sorrow, & increase the joy of the minde.

The leaves and floures of Borrage put into wine make men and women glad and merry, driving away all sadnesse, dulnesse, and melancholy.

Syrrup made of the floures of Borrage comforteth the heart, purgeth melancholy, and quieteth the phrenticke or lunaticke person.

Syrrup made of the juice of Borrage with sugar, adding thereto pouder of the bone of a Stags heart, is good against swouning, the cardiacke passion of the heart, against melancholy and the falling sicknesse.

Alkanet or wilde Buglosse

These herbes comprehended under the name of *Anchusa*, were so called of the Greeke word, that is, to colour or paint any thing: Whereupon those plants were called *Anchusa*, of that flourishing and bright red colour which is in the root, even as red as pure and cleare bloud.

Alkanet hath many leaves like *Echium* or small Buglosse, covered over with a pricky hoarinesse, having commonly but one stalke, which is round, rough, and a cubit high. The cups of the floures are of a sky colour tending to purple: the seed is small, somwhat long, and of a pale colour: the root is a finger thicke, the pith or inner part thereof is of a wooddy substance, dying the hands or whatsoever toucheth the same, of a bloudy colour, or of the colour of Sanders.

Divers of the later Physitions do boile with the root of Alkanet and wine, sweet butter, such as hath in it no salt at all, untill such time as it becommeth red, which they call red butter, and give it not only to those that have falne from some high place, but also report it to be good to drive forth the measels and small pox, if it be drunke in the beginning with hot beere.

The roots of these are used to color sirrups, waters, gellies, & such like infections as Turnsole is.

John of *Ardern* hath set down a composition called *Sanguis Veneris*, which is most singular in deep punctures or wounds made with thrusts, as follows: take of oile olive a pint, the root of Alkanet two ounces, earth worms purged, in number twenty, boile them together & keep it to the use aforesaid.

The Gentlewomen of France do paint their faces with these roots, as it is said.

TARRAGON

Tarragon the sallade herbe hath long and narrow leaves of a deep green colour, greater and longer than those of common Hyssope, with slender brittle round stalkes two cubites high: about the branches whereof hang little round floures, never perfectly opened, of a yellow colour mixed with blacke, like those of common Wormewood.

Tarragon is cherished in gardens, and is encreased by the young shoots: *Ruellius* and such others have reported many strange tales hereof scarce worth the noting, saying, that the seed of flax put into a Raddish root or sea Onion, and so set, doth bring forth this herbe Tarragon.

It is greene all Summer long, and a great part of Autumne, and floureth in July.

It is called in Latine, *Draco*; in French, *Dragon*; in English, Tarragon. *Simeon Sethi* the Greeke also maketh mention of *Tarchon*.

Tarragon is not to be eaten alone in sallades, but joyned with other herbs, as Lettuce, Purslain, and such like, that it may also temper the coldnesse of them, like as Rocket doth, neither do we know what other use this herbe hath.

INDIAN CRESSES

Cresses of India have many weake and feeble branches, rising immediatly from the ground, dispersing themselves far abroad; by meanes whereof one plant doth occupie a great circuit of

Tarragon

ground, as doth the great Bindeweede. The tender stalks divide themselves into sundry branches, trailing likewise upon the ground, somewhat bunched or swollen up at every joint or knee, which are in colour of a light red, but the spaces betweene the joints are greene. The leaves are round like wall peniwort, called Cotyledon, the foot-stalke of the leafe commeth forth on the backe-side almost in the middest of the leafe, as those of Frogbit, in taste and smell like the garden Cresses. The flours are dispersed throughout the whole plant, of colour yellow, with a crossed star overthwart the inside, of a deepe Orange colour: unto the backe-part of the same doth hang a taile or spurre, such as hath the Larkes heele, called in Latine *Consolida Regalis*; but greater, and the spur or heele longer; which beeing past there succeed bunched and knobbed coddes or seed vessells, wherein is contained the seed, rough, browne of colour, and like unto the seeds of the beete, but smaller.

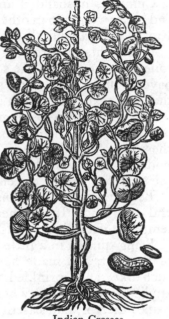

The seeds of this rare and faire plant came from the Indies into Spaine, and thence into France and Flanders, from whence I received seed that bore with mee both floures & seed, especially those I received from my loving friend *John Robin* of Paris.

This beautifull Plant is called in Latine, *Nasturtium Indicum*: in English, Indian Cresses. Although some have deemed it a kind of *Convolvulus*, or Binde-weed; yet I

Indian Cresses

am well contented that it retaine the former name, for that the smell and taste show it to be a kinde of Cresses.

CUMIN

This garden Cumin is a low or base herbe of a foot high: the stalke divideth it self into divers small branches, whereon doe grow little jagged leaves very finely cut into small parcels, like those of Fennell, but more finely cut, shorter and lesser, the spoky tufts grow at the top of the branches and stalkes, of a red or purplish colour: after which come the seed, of a strong or rancke smell, and biting taste: the root is slender, which perisheth when it hath ripened his seed.

Cumin is husbanded and sowne in Italy and Spaine, and is very common in other hot countries, as in Æthiopia, Egypt, Cilicia, and all the lesser Asia.

It delights to grow especially in putrified and hot soiles: I have proved the seeds in my garden, where they have brought forth ripe seed much fairer and greater than any that comes from beyond the seas.

It is to be sown in the middle of the spring; a showre of rain presently following much hindreth the growth thereof, as *Ruellius* saith. My self did sow it in the midst of May, which sprung up in six daies after: and the seed was ripe in the end of July.

Being taken in a supping broth it is good for the chest and cold lungs. It stancheth bleeding at the nose, being tempered with vineger and smelt unto.

Being quilted in a little bag with some small quantitie of Bay salt, and made hot upon a bed-pan with fire or such like, and sprinkled with good wine vineger, and applied to the side very hot, it taketh away the stitch and paines thereof, and easeth the pleurisie very much.

Water Saligot, water Caltrops, or water Nuts

Water Caltrops have long slender stalks growing up and rising from the bottom of the water, and mounting above the same: the root is long, having here & there under the water certaine tassels full of small strings or threddy haires: the stem towards the top of the water is very great in respect of that which is lower; the leaves are large and somewhat round, not unlike those of the Poplar or Elme tree leaves, a little crevised or notched about the edges: amongst or under the leaves grow the fruit, which is triangled, hard, sharp pointed and prickly, in shape like those hurtfull engins in the wars, cast in the passage of the enemy to annoy the feet of their horses, called Caltrops, whereof this tooke it's name: within these heads or Nuts is contained a white kernell in taste almost like the Chesnut, which is reported to bee eaten green, and being dried and ground to serve in stead of bread.

Cordus saith that it groweth in Germanie in myrie lakes, and in city ditches that have mud in them: in Brabant and other places of the Low countries it is found often times in standing waters and springs.

SEA HOLLY

Water Caltrops

Sea Holly hath broad leaves almost like to Mallow leaves, but cornered in the edges, and set round about with hard prickles, fat, of a blewish white, and of an aromatical or spicy taste: the stalke is thick, about a cubit high, now and then somwhat red below: it breaketh forth in the tops into prickly round heads or knops, of the bignesse of a Wall-nut, held in for the most part with six prickly leaves compassing the top of the stalke round about; which leaves as well as the heads are of a glistering blew: the floures forth of the heads are likewise blew, with white threds in the midst: the root is of the bignesse of a mans finger, so very long, as that it cannot be all plucked up but very seldome; set here and there with knots, and of taste sweet and pleasant.

Eryngium marinum growes by the sea side upon the baich and stony ground. I found it growing plentifully at Whitstable in Kent, at Rie and Winchelsea in Sussex, and in Essex at Landamer lading, at Harwich, and upon Langtree point on the other side of the water, from whence I brought plants for my garden.

The roots condited or preserved with sugar as hereafter followeth, are exceeding good to be given to old and aged people that are consumed and withered with age, and which want natural moisture: they are also good for other sorts of people that have no delight, nourishing and restoring the aged, and amending the defects of nature in the yonger.

¶ *The manner to condite Eringos.*

Refine sugar fit for the purpose, and take a pound of it, the white of an egge, and a pinte of cleer water, boile them together and scum it, then let it boile until it be come to good strong syrrup, and when it is boiled, as it cooleth adde thereto a saucer full of rose water, a spoone full of Cinnamon water, and a grain of muske, which have been infused together the night before, and now strained: into which syrrup being more than halfe cold put in your roots to soke and infuse untill the next day; your roots being ordered in manner hereafter following:

These your roots being washed and picked, must be boiled in faire water by the space of foure houres, til they be soft: then must they be pilled clean as ye pil parsneps, & the pith must be drawn out at the end of the root: but if there be any whose pith cannot be drawn out at the end, then you must slit them and so take it out: these you must also keep from much handling, that they may be clean: let them remain in the syrrup till the next day, and then set them on the fire in a faire broad pan untill they be very hot, but let them not boile at all: let them remain over the fire an houre or more, remooving them easily in the pan from one place to another with a wooden slice. This done, have in a readinesse great cap or royall papers, whereupon strow some sugar, upon which lay your roots, having taken them out of the pan. These papers you must put into a stouve or hot-house to harden; but if you have not such a place, lay them before a good fire: in this maner if you condite your roots, there is not any that can prescribe you a better way. And thus you may condite any other root whatsoever, which will not only be exceeding delicat, but very wholsome, and effectuall against the diseases above named.

They report of the herb sea Holly, if one goat take it into her mouth, it causeth her first to stand still, and afterwards the whole flocke, untill such time as the sheep-heard take it from her mouth. *Plutarch.*

TEASELS

Garden Teasell bringeth forth a stalke that is straight, very long, jointed, and ful of prickles: the leaves grow forth of the joynts by couples, not onely opposite or set one right against another, but also compassing the stalke about, and fastened together; and so fastened, that they hold dew and raine water in manner of a little bason: these be long, of a light greene colour, and like to those of Lettice, but full of prickles in the edges, and have on the outside all alongst the ridge stiffer prickles: on the tops of the stalkes stand heads with sharpe prickles like those of the Hedge-hog, and crooking backward at the point like hookes: out of which heads grow little floures: The seed is like Fennell-seed, and in taste bitter: the heads wax white when they grow old, and there are found in the midst of them when they are cut, certaine little magots: the root is white, and of a meane length.

The tame Teasell is sowne in this country in gardens, to serve the use of Fullers and Clothworkers.

Teasell is called in Latine, *Dipsacus*, and *Laver Lavacrum*, of the forme of the leaves made up in fashion of a bason, which is never without water: in English, Teasell, Carde Teasell, and Venus bason.

Wilde Teasell

It is needlesse here to alledge those things that are added touching the little wormes or magots found in the heads of the Teasell, and which are to be hanged about the necke, or to mention the like thing that *Pliny* reporteth of Galedragon: for they are nothing else but most vaine and trifling toies, as my selfe have proved a little before the impression hereof, having a most grievous ague, and of long continuance : notwithstanding Physicke charmes, these worms hanged about my neck, spiders put into a walnut shell, and divers such foolish toies that I was constrained to take by fantasticke peoples procurement; notwithstanding, I say, my helpe came from God himselfe, for these medicines and all other such things did me no good at all.

RUE, OR HERBE GRACE

Garden Rue is a shrub full of branches, now and then a yard high, or higher: the stalkes whereof are covered with a whitish barke, the branches are more green: the leaves hereof consist of divers parts, and be divided into wings, about which are certaine little ones, of an odde number, something broad, more long than round, smooth and somewhat fat, of a gray colour or greenish blew: the floures in the top of the branches are of a pale yellow consisting of foure little leaves, something hollow: in the middle of which standeth up a little head or button foure square, seldome five square, containing as many little coffers as it hath corners, being compassed about with divers little yellow threds: out of which hang pretty fine tips of one colour; the seed groweth in the little coffers: the root is wooddy, and fastned with many strings: this Rue hath a very strong and ranke smell, and a biting taste.

Pliny saith that there is such friendship between it and the fig-tree, that it prospers no where so well as under the fig tree. The best for physicks use is that which groweth under the fig tree, as *Dioscorides* saith: the cause is alledged by *Plutarch, lib.* 1. of his *Symposiacks* or Feasts, for he saith it becomes more sweet and milde in taste, by reason it taketh as it were some part of the sweetnesse of the fig tree, whereby the over-ranke qualitie of the Rue is allayed; unlesse it be that the figge tree whilest it drawes nourishment to it selfe, draweth also the ranknesse away from the Rue.

The herb a little boiled or skalded, and kept in pickle as Sampier, and eaten, quickens the sight. The same applied with hony and the juice of Fennell, is a remedie against dim eies.

The juice of Rue made hot in the rinde of a pomegranat and dropped into the eares, takes away the pain of thereof.

Dioscorides saith, That Rue put up in the nosthrils stayeth bleeding. So saith *Pliny* also; when notwithstanding it is of power rather to procure bleeding, through its sharpe and biting quality.

Dioscorides writeth, That a twelve penny weight of the seed drunke in wine is a counterpoison against deadly medicines or the poison of Wolfes-bane, Mushroms or Toad-stooles, the biting of Serpents, the stinging of Scorpions, Bees, hornets, and wasps; and is reported, That if a man bee anointed with the juice of Rue, these will not hurt him; and that the serpent is driven away at the smell thereof when it is burned: insomuch that when the Weesell is to fight with the serpent, shee armeth her selfe by eating Rue, against the might of the Serpent.

The leaves of Rue eaten with the kernels of Walnuts or figs stamped together and made into a masse or paste, is good against all evill aires, the pestilence or plague, resists poison and all venome.

Ruta sylvestris or wild Rue is more vehement both in smel and operation, and therefore the more virulent or pernitious; for sometimes it fumeth out a vapor or aire so hurtfull that it scorches the face of him that looketh upon it, raising up blisters, wheals, and other accidents: it venometh their hands that touch it, and will infect the face also if it be touched before they be clean washed: wherfore it is not to be admitted to meat or medicine.

ONIONS

The Onion hath narrow leaves, and hollow within; the stalke is single, round, biggest in the middle, on the top whereof groweth a round head covered with a thin skin or film, which being broken, there appeare little white floures made up in form of a ball, and afterward blacke seed three cornered, wrapped in thin white skins. In stead of the root there is a bulbe or round head compact of many coats, which often times becommeth great in manner of a Turnep, many times long like an egge. To be briefe, it is covered with very fine skins for the most part of a whitish colour.

The Onion requireth a fat ground well digged and dunged, as *Palladius* saith. It is cherished everie where in kitchen gardens, now and then sowne alone, and many times mixed with other herbs.

The Onions do bite, attenuate or make thin, and cause drynesse: being boiled they do lose their sharpnesse, especially if the water be twice or thrice changed, and yet for all that they doe not lose their attenuating qualitie.

The juice of Onions snuffed up into the nose, purgeth the head, and draweth forth raw flegmaticke humors. Stamped with Salt, Rue, and Honey, and so applied, they are good against the biting of a mad Dog. Rosted in the embers and applied, they ripen and breake cold Apostumes, Biles, and such like.

The juice of Onions mixed with the decoction of Penni-royall, and anointed upon the goutie member with a feather, or a cloath wet therein, and applied, easeth the same very much. The juice anointed upon a pild or bald head in the Sun, bringeth the haire againe very speedily.

The juice taketh away the

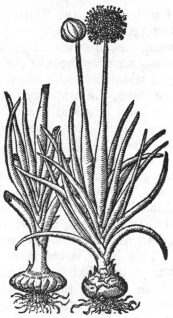

White Onions

heat of scalding with water or oile, as also burning with fire & gunpouder, as is set forth by a very skilfull Surgeon Mr. *William Clowes* one of the Queens Surgeon; and before him by *Ambrose Parey*, in his treatise of wounds made by gun-shot.

Onions sliced and dipped in the juice of Sorrell, and

177

given unto the Sicke of a tertian Ague, to eat, takes away the fit in once or twice so taking them.

¶ The Onion being eaten, yea though it be boiled, causeth head-ache, hurteth the eyes, and maketh a man dim sighted, dulleth the sences, and provoketh over-much sleep, especially being eaten raw.

SKIRRETS

The leaves of the Skirret consist of many small leaves fastened to one rib, every particular one whereof is some-thing nicked in the edges, but they are lesser, greener, & smoother than those of the Parsnep. The stalkes be short, and seldome a cubit high; the floures in the spokie tufts are white, the roots bee many in number, growing out of one head an hand breadth long, most commonly not a finger thick, they are sweet, white, good to be eaten, and most pleasant in taste.

This skirret is planted in Gardens, and especially by the root, for the greater and thicker ones being taken away, the lesser are put into the earth againe: which thing is best to be done in March or Aprill, before the stalkes come up, and at this time the roots which bee gathered are eaten raw, or boyled.

This herb is called in Latine, *Sisarum*, in English, Skirret and Skirwort. And this is that *Siser* or Skirret which *Tiberius* the Emperour commanded to bee con-veied unto him from Gelduba a castle about the river of Rhene, as *Pliny* reporteth in *lib*. 16. *cap*. 5. The Skirret is a medicinable herbe, and is the same that the foresaid Emperour did so much commend, insomuch that he desired the same to be brought unto him every yeare out of Germany.

The roots be eaten boiled, with vineger, salt, and a little oyle, after the manner of a sallad, and oftentimes they be fried in oile and butter, and also dressed after other fashions, according to the skill of the cooke, and the taste of the eater.

FENNELL

The Fennell, called in Latine, *Fœniculum*, is so well knowne amongst us, that it were but lost labour to describe the same.

Common Fennell

The second kinde of Fennell is likewise well knowne by the name of Sweet Fennell, so called because the seeds thereof are in taste sweet like unto Annise seeds, resembling the common Fennell, saving that the leaves are larger and fatter, or more oleous: the seed greater and whiter, and the whole plant in each respect greater.

These herbs are set and sowne in gardens.

The pouder of the seed of Fennell drunke for certaine daies together fasting preserveth the eye-sight: whereof was written this Distichon following:

Of Fennell, Roses, Vervain, Rue, and Celandine,
Is made a water good to cleere the sight of eine.

PARSNEPS

There is a good and pleasant food or bread made of the roots of Parsneps, as my friend Mr. *Plat* hath set forth in his booke of experiments, which I have made no tryall of, nor meane to do.

CUCUMBERS

The Cucumber creepes alongst upon the ground all about, with long rough branches; whereupon doe grow broad rough leaves uneven about the edges: from the bosome whereof come forth crooked clasping tendrels like those of the Vine. The floures shoot forth betweene the stalkes and the leaves, set upon tender footstalkes composed of five small yellow leaves: which being past, the fruit succeedeth, long, cornered, rough, and set with certaine bumpes or risings, greene at the first, and yellow when they be ripe, wherein is contained a firme and sollid pulpe or substance transparent or thorow-shining, which together with the seed is eaten a little before they be fully ripe. The seeds be white, long, and flat.

There be also certaine long cucumbers, which were first made (as is said) by art and manuring, which Nature afterwards did preserve: for at the first, when as the fruit is very little, it is put into some hollow cane, or other thing made of purpose, in which the cucumber groweth very long, by reason of that narrow hollownesse, which being filled up, the cucumber encreaseth in length. The seeds of this kinde of cucumber being sowne bringeth forth not such as were before, but such as art hath framed; which of their own growth are found long, and oftentimes very crookedly turned: and thereupon they have beene called *Anguini*, or long Cucumbers.

The Cucumber is named generally *Cucumis*: in shops, *Cucumer*: in English, Cowcumbers and Cucumbers.

The fruit cut in pieces or chopped as herbes to the pot, and boiled in a small pipkin with a piece of mutton, being made into potage with Ote-meale, even as herb potage are made, whereof a messe eaten to break-fast, as much to dinner, and the like to supper; taken in this manner for the space of three weekes together without intermission, doth perfectly cure all manner of sauce flegme and copper faces, red and shining fierie noses (as red as red Roses) with pimples, pumples, rubies, and such like precious faces.

Provided alwaies that during the time of curing you doe use to wash or bathe the face with this liquor following.

Take a pinte of strong white wine vinegre, pouder of the roots of Ireos or Orrice three dragmes, searced or bolted into most fine dust, Brimmestone in fine pouder halfe a ounce, Camphire two dragmes, stamped with two blanched Almonds, foure Oke Apples cut thorow the middle, and the juyce of foure Limons: put them all together in a strong double glasse, shake them together very strongly, setting the same in the Sunne for the space of ten daies: with which let the face be washed and bathed daily, suffering it to drie of it selfe without wiping it away. This doth not onely helpe fierie faces, but also taketh away lentils, spots, morphew, Sun-burne, and all other deformities of the face.

¶ I have thought it good and convenient in this place to set downe not onely the time of sowing and setting of Cucumbers, Muske-melons, Citruls, Pompions, Gourds, and such like, but also how to set or sow all manner and kindes of other colde seeds, as also whatsoever strange seeds are brought unto us from the Indies, or other hot Regions: *videl.*

First of all in the middest of Aprill or somewhat sooner (if the weather be anything temperate) you shall cause to be made a bed or banke of hot and new horsedung taken from the stable (and not from the dunghill) of an ell in breadth, and the like in depth or thicknesse, of what length you please, according to the quantity of your seed: the which bank you shall cover with hoops or poles, that you may the more conveniently cover the whole bed or banke with Mats, old painted cloth, straw or such like, to keepe it from the injurie of the cold frosty nights, and

not hurt the things planted in the bed: then shall you cover
the bed all over with the most fertilest earth finely sifted,
halfe a foot thicke, wherein you shall set or sow your
seeds: that being done, cast your straw or other coverture
over the same; and so let it rest without looking upon it, or
taking away of your covering for the space of seven or
eight daies at the most, for commonly in that space they
will thrust themselves up nakedly forth of the ground:
then must you cast upon them in the hottest time of the
day some water that hath stood in the house or in the Sun
a day before, because the water so cast upon them newly
taken forth of a well or pumpe, will so chill and coole them
being brought and nourished up in such a hot place, that
presently in one day you have lost all your labour; I mean
not only your seed, but your banke also; for in this space
the great heate of the dung is lost and spent, keeping in
memory that every night they must be covered and opened
when the day is warmed with the Sun beames: this must
be done from time to time untill that the plants have foure
or six leaves a piece, and that the danger of the cold nights
is past: then must they be replanted very curiously with
the earth sticking to the plant, as neere as may be unto the
most fruitfull place, and where the Sun hath most force in
the garden; provided that upon the removing of them you
must cover them with some Docke leaves or wispes of
straw, propped up with forked stickes, as well to keepe
them from the cold of the night, as also the heat of the
Sun: for they cannot whilest they be young and newly
planted, endure neither overmuch cold nor overmuch
heate, untill they are well rooted in their new place or
dwelling.

Oftentimes it falleth out that some seeds are more
franker and forwarder than the rest, which commonly do
rise up very nakedly with long necks not unlike to the
stalke of a small mushrome, of a night old. This naked
stalke must you cover with the like fine earth even to the

greene leaves, having regard to place your banke so that it may be defended from the North-windes.

Observe these instructions diligently, and then you shall not have cause to complaine that your seeds were not good, nor of the intemperancie of the climat (by reason wherof you can get no fruit) although it were in the furthest parts of the North of Scotland.

M U S K E - M E L O N , O R M I L L I O N

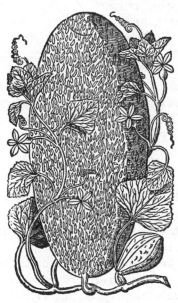

Spanish Melon

That which the later Herbarists do call Muske-Melons is like to the common Cucumber in stalks, lying flat upon the ground, long branched, and rough. The leaves be much alike, yet are they lesser, rounder, and not so cornered: the floures in like manner bee yellow; the fruit is bigger, at the first somwhat hairy, somthing long, now and then somwhat round; oftentimes greater, and many times lesser: the barke or rinde is of an overworne russet greene colour, ribbed and furrowed very deepely, having often chaps or chinks, and a confused roughnesse: the pulp or inner substance which is to be eaten is of a feint yellow colour; the middle part whereof is full of a slimie moisture: amongst which is contained the seed, like of those of the Cucumber, but lesser, and of a browner colour.

The sugar Melon hath long trailing stalks lying upon

183

the ground, whereon are set small clasping tendrels like those of the Vine, and also leaves like unto the common Cucumber, but of a greener colour: the fruit commeth forth among those leaves, standing upon slender foot-stalkes, round as the fruit of Coloquintida, and of the same bignesse, of a most pleasant taste like sugar, whereof it tooke the syrname *Saccharatus*.

The Spanish Melon brings forth long trailing branches, wheron are set broad leaves slightly indented about the edges, not divided at all, as are all the rest of the Melons. The fruit groweth neere unto the stalke.

They delight in hot regions, notwithstanding I have seen at the Queens house at S. *James* many of the first sort ripe, through the diligent and curious nourishing of them by a skilfull gentleman the keeper of the said house called Mr. *Fowle*: and in other places neere the right honourable Lord of Sussex his house of Bermondsey by London, where yearely there is very great plenty, especially if the weather be any thing temperat.

GOURDS

There are divers sorts of Gourds, some wilde, others tame of the garden: some bearing fruit like unto a bottle; others long, bigger at the end, keeping no certain form or fashion; some greater, others lesse.

The Gourd bringeth forth very long stalks as be those of the Vine, cornered and parted into divers branches, which with his clasping tendrels taketh hold and clymeth upon such things as stand neere unto it: the leaves bee very great, broad, and sharpe pointed, almost as great as those of the Clot-burre, but softer, and somwhat covered as it were with a white freese, as be also the stalkes and branches, like those of the marish Mallow: the floures be white, and grow forth from the bosome of the leaves: in their places come up the fruit, which are not all of one fashion, for oftentimes they have the forme of flagons and bottles, with a great large belly and a small necke. The Gourd (saith *Pliny, lib.* 19. *cap.* 5.) groweth into any forme or fashion that you would have it; either like unto a wreathed Dragon, the leg of a man, or any other shape, according to the mold wherein it is put while it is yong: being suffered to clyme upon any arbor where the fruit may hang, it hath bin seen to be nine foot long, by reason of his great weight which hath stretched it out in the length: the rinde when it is ripe, is very hard, woody, and of a yellow colour: the meat or inward pulpe is white; the seed long, flat, pointed at the top, broad, below, with two peaks standing out like hornes, white within, and sweet of taste.

Gourds are cherished in the gardens of these cold regions rather for pleasure than profit: in the hot countries where they come to ripenes they are somtimes eaten, but with small delight; especially they are kept for the rinds, wherein they put turpentine, oile, hony, and also serve them as pales to fetch water in, and many other the like uses.

A long Gourd or Cucumber being laid in the cradle or bed by the yong infant whilest it is asleep and sicke of an ague, it shall be very quickly made whole.

The Marvell of the World

This admirable Plant, called the Marvell of Peru, or the Marvell of the World, springs forth of the ground like unto Basil in leaves; among which it sendeth out a stalke two cubits and a halfe high, of the thicknesse of a finger, full of juice, very firme, and of a yellowish green colour, knotted or kneed with joints somewhat bunching forth, of purplish colour, as in the female Balsamina: which stalke divideth it selfe into sundry branches or boughes, and those also knottie like the stalke. His branches are decked with leaves growing by couples at the joints like the leaves of wilde Peascods, greene, fleshy, and full of joints; which being rubbed doe yeeld the like unpleasant smell as wilde Peascods do, and are in taste also very unsavory, yet in the later end they leave a tast and sharp smack of Tabaco. The stalks toward the top are garnished

The Marvell of Peru

with long hollow single floures, folded as it were into five parts before they be opened; but being fully blown, do resemble the floures of Tabaco, not ending in sharp corners, but blunt & round as the flours of Bindweed, and larger than the floures of Tabaco, glittering oft times with a fine purple or crimson colour, many times of an horse-flesh, sometimes yellow, sometimes pale, and somtime resembling an old red or yellow colour; sometime whitish, and most commonly two colours occupying half the floure, or intercoursing the whole floure with streaks or orderly streames, now yellow, now purple, divided through the whole, having sometime great, somtime little spots of a purple colour, sprinkled and scattered in a most variable order and brave mixture. The ground or field of the whole floure is either pale, red, yellow, or white, containing in the middle of the hollownesse a pricke or pointal set round about with six small strings or chives. The floures are very sweet and pleasant, resembling the Narcisse or white Daffodill, and are very suddenly fading; for at night they are floured wide open, and so continue untill eight of the clocke the next morning, at which time they begin to close (after the maner of Bindweed) especially if the weather be very hot: but the aire being temperat, they remain open the whole day, and are closed only at night, and so perish, one floure lasting but onely one day, like the true Ephemerum or Hemerocallis.

This marvellous variety doth not without cause bring admiration to all that observe it. For if the floures be gathered and reserved in severall papers, and compared with those floures that will spring and flourish the next day, you shall easily perceive that one is not like another in colour, though you shall compare one hundred which floure one day, and another hundred which you gather the next day, and so from day to day during the time of their flouring. It bringeth new floures from July unto October in infinite number, yea even untill the frosts doe cause the whole plant to perish: notwithstanding it may be reserved in pots, and set in chambers and cellars that are warme, and so defended from the injurie of our cold climate; provided alwaies that there be not any water cast upon the pot, or set forth to take any moisture in the aire untill March following; at which time it must be taken forth of the pot and replanted in the garden. By this meanes I have preserved many (though to small purpose) because I have sowne seeds that have borne floures in as ample manner and in as good time as those reserved plants.

Of this wonderfull herbe there be other sorts, but not so amiable or so full of varietie, and for the most part their floures are all of one color. But I have since by practise found out another way to keepe the roots for the yere following with very little difficultie, which never faileth. At the first frost I dig up the roots and put up or rather hide the roots in a butter ferkin, or such like vessell, filled with the sand of a river, the which I suffer still to stand in some corner of an house where it never receiveth moisture untill Aprill or the midst of March, if the weather be warme; at which time I take it from the sand and plant it in the garden, where it doth flourish exceeding well and increaseth by roots; which that doth not which was either sowne of seed the same yeere, nor those plants that were preserved after the other manner.

MADDE APPLES

Raging Apples hath a round stalke of two foot high, divided into sundry branches, set with broad leaves somewhat indented about the edges, not unlike the leaves of white Henbane, of a darke browne greene colour, somewhat rough. Among the which come the floures of a white colour, and somtimes changing into purple, made of six parts wide open like a star, with certain yellow chives or thrums in the middle: which being past, the fruit comes in place, set in a cornered cup or huske after the manner of great Nightshade, great and somewhat long, of the bignesse of a Swans egge, and sometimes much greater, of a white colour, somtimes yellow, and often brown, wherein is contained small flat seed of a yellow colour. The root is thick, with many threds fastned thereto.

This Plant growes in Egypt almost every where in sandy fields even of it selfe, bringing forth fruit of the bignesse of a great Cucumber, as *Petrus Bellonius* writeth, *lib*. 2. of his singular observations.

We had the same in our London gardens, where it hath borne floures; but Winter approaching before the time of ripening, it perished: neverthelesse it came to beare fruit of the bignes of a goose egg one extraordinarie temperate yeare, as I did see in the garden of a worshipfull merchant Mr. *Harvy* in Limestreet; but never to the full ripenesse.

The people of Toledo eat them with great devotion, being boiled with fat flesh, putting to it some scraped cheese, which they do keep in vineger, hony, or salt pickle all winter.

Petrus Bellonius and *Hermolaus Barbarus* report, That in Egypt & Barbary they use to eat the fruit of *Mala insana* boiled or rosted under ashes, with oile, vineger, and pepper, as people use to eat Mushroms. But I rather wish English men to content themselves with the meat and sauce of our owne country, than with fruit and sauce eaten with such perill; for doubtlesse these Apples have a mischievous qualitie, the use whereof is utterly to bee forsaken. As wee see and know many have eaten and do eat Mushroms more for wantonnesee than for need; for there are two kinds therof deadly, which being dressed by an unskilfull cooke may procure untimely death: it is therefore better to esteem this plant and have it in the garden for your pleasure and the rarenesse thereof, than for any vertue or good qualities yet knowne.

APPLES OF LOVE

The Apple of Love bringeth forth very long round stalkes or branches, fat and full of juice, trailing upon the ground, not able to sustain himselfe upright by reason of the tendernesse of the stalkes, and also the great weight of the leaves and fruit wherewith it is surcharged. The leaves are great, and deeply cut or jagged about the edges, not unlike to the leaves of Agrimonie, but greater, and of a whiter greene colour: Amongst which come forth yellow floures growing upon short stems or footstalkes, clustering together in bunches: which being fallen there doe come in place faire and goodly apples, chamfered, uneven, and bunched out in many places; of a bright shining red colour, and the bignesse of a goose egge or a large pippin. The pulpe or meat is very full of moisture, soft, reddish, and of the substance of a wheat plumme. The seed is small, flat and rough: the root small and threddy: the whole plant is of a ranke and stinking savour.

There hath happened unto my hands another sort, agreeing very notably with the former, as well in leaves and stalkes as also in floures and roots, onely the fruit hereof was yellow of colour, wherein consisteth the difference.

Apples of Love grow in Spaine, Italie, and such hot Countries, from whence my selfe have received seeds for my garden, where they doe increase and prosper.

It is sowne in the beginning of Aprill in a bed of hot horse-dung, after the maner of muske Melons and such like cold fruits.

The Apple of Love is called in Latine *Pomum Aureum*, *Poma Amoris*, and *Lycopersi-cum*: of some, *Glaucium*: in English, Apples of Love, and Golden Apples: in French, *Pommes d'amours*. Howbeit there be other golden Apples whereof the Poëts doe fable, growing in the Gardens of the

Apples of Love

daughters of *Hesperus*, which a Dragon was appointed to keepe, who, as they fable, was killed by *Hercules*.

The Golden Apple, with the whole herbe it selfe is cold, yet not fully so cold as Mandrake, after the opinion of *Dodonæus*. But in my judgement it is very cold, yea perhaps in the highest degree of coldnesse: my reason is, because I have in the hottest time of Summer cut away

190

the superfluous branches from the mother root, and cast
them away carelesly in the allies of my Garden, the which
(notwithstanding the extreme heate of the Sun, the hardnesse of the trodden allies, and at that time when no rain
at all did fal) have growne as fresh where I cast them, as
before I did cut them off; which argueth the great coldnesse contained therein. True it is, that it doth argue also a
great moisture wherewith the plant is possessed, but as I
have said, not without great cold, which I leave to every
mans censure.

In Spaine and those hot Regions they use to eate the
Apples prepared and boiled with pepper, salt, and oyle:
but they yeeld very little nourishment to the body, and
the same naught and corrupt.

Likewise they doe eate the Apples with oile, vinegre
and pepper mixed together for sauce to their meat, even
as we in these cold countries doe Mustard.

THORNIE APPLES

The stalkes of Thorny-apples are oftentimes above a cubit
and a halfe high, seldome higher, an inch thicke, upright
and straight, having very few branches, sometimes none
at all, but one upright stemme: whereupon doe grow
leaves smooth and even, little or nothing indented about
the edges, longer and broader than the leaves of Nightshade, or of the mad Apples. The floures come forth of
long toothed cups, great, white, of the forme of a bell, or
like the floures of the great Withwinde that rampeth in
hedges; but altogether greater and wider in the mouth,
sharpe cornered at the brimmes, with certaine white
chives or threds in the middest, of a strong ponticke
savour, offending the head when it is smelled unto: in
the place of the floure commeth up round fruit full of
short and blunt prickles of the bignesse of a green Wall-
nut when it is at the biggest, in which are the seeds of the

bignesse of tares or of the seed of Mandrakes, and of the same forme. The herbe it selfe is of a strong savor, and doth stuffe the head, and causeth drowsinesse. The root is small and threddy.

There is another kinde hereof altogether greater than the former, whose seeds I received of the right honourable the Lord *Edward Zouch*; which he brought from Constantinople, and of his liberalitie did bestow them upon me, as also many other rare and strange seeds; and it is that Thorn-apple that I have dispersed through this land, whereof at this present I have great use in Surgery; as well in burnings and scaldings, as also in virulent and maligne ulcers, apostumes, and such like. The which plant hath a very great stalke in fertile ground, bigger than a man's arme, smooth and greene of colour, which a little above the ground divideth it selfe into sundry branches or armes in manner of an hedge tree; whereupon are placed many great leaves cut and indented deepely about the edges, with many uneven sharpe corners: among these leaves come white round floures made of one piece in manner of a bell, shutting it selfe up close toward night, as doe the floures of the great Binde-weed, whereunto it is very like, of a sweet smell, but so strong, that it offends the sences. The fruit followeth round, sometimes of the fashion of an egge, set about on every part with most sharpe prickles; wherein is contained very much seed of the bignesse of tares, and of the same fashion. The root is thicke, made of great and small strings: this plant is sowen, beareth his fruit, and perisheth the same yeare.

The juice of Thorn-apples boiled with hogs grease to the form of an unguent or salve, cures all inflammations whatsoever, all manner of burnings or scaldings, as well of fire, water, boiling lead, gun-pouder, as that which comes by lightning, and that in very short time, as my selfe have found by my daily practise, to my great credit and profit. The first experience came from Colchester, where Mistresse *Lobel* a merchants wife there being most grievously burned by lightning, and not finding ease or cure in any other thing, by this found helpe and was perfectly cured when all hope was past, by the report of Mr. *William Ram* publique Notarie of the said towne.

The leaves stamped small and boiled with oile Olive untill the herbs be as it were burnt, then strained and set to the fire again, with some wax, rosin, and a little turpentine, and made into a salve, doth most speedily cure new and fresh wounds.

BITTER-SWEET, OR WOODDY NIGHTSHADE

Bitter-sweet

Bitter-sweet bringeth forth wooddy stalks as doth the Vine, parted into many slender creeping branches, by which it climeth and taketh hold of hedges and shrubs next unto it. The barke of the oldest stalks are rough and whitish, of the colour of ashes, with the outward rind of a bright green colour, but the yonger branches are green as are the leaves: the wood brittle, having in it a spongie pith: it is clad with long leaves, smooth, sharp pointed, lesser than those of the Bindweed. At the lower part of the same leaves doth grow on either side one smal or lesser leafe like unto two eares. The floures be small, and somewhat clustered together, consisting of

193

five little leaves apiece of a perfect blew colour, with a certain pricke or yellow pointal in the middle: which being past, there do come in place faire berries more long than round, at the first green, but very red when they be ripe; of a sweet taste at the first, but after very unpleasant, of a strong savor, growing together in clusters like burnished coral. The root is of a mean bignesse, and full of strings.

I have found another sort which bringeth forth most pleasant white flours, with yellow pointals in the middle: in other respects agreeing with the former.

Bitter-sweet growes in moist places about ditches, rivers, and hedges, almost everie where.

The other sort with the white floures I found in a ditch side, against the right honourable the Earle of Sussex his garden wall, at his house in Bermondsey street by London, as you go from the court which is full of trees, unto a ferm house neere thereunto.

The leaves come forth in the spring, the flours in July, the berries are ripe in August.

The later Herbarists have named this plant *Dulcamara*, *Amaro dulcis*, & *Amaradulcis*; *Pliny* calleth it *Melortum*: *Theophrastus*, *Vitis sylvestris*: in English we call it Bitter-sweet, and wooddy Nightshade. But every Author must for his credit say something, although but to smal purpose; for *Vitis sylvestris* is that which wee call our Ladies Seale, which is no kinde of Nightshade.

MANDRAKE

The male Mandrake hath great broad long smooth leaves of a darke greene colour, flat spred upon the ground: among which come up the floures of a pale whitish colour, standing every one upon a single small and weake foot-stalke of a whitish greene colour: in their places grow round Apples of a yellowish colour, smooth, soft, and glittering, of a strong smell: in which are contained flat and smooth seeds in fashion of a little kidney, like those of the Thorne-apple. The root is long, thicke, whitish, divided many times into two or three parts resembling the legs of a man, as it hath been reported; whereas in truth it is no otherwise than in the roots of carrots, parseneps, and such like, forked or divided into two or more parts, which Nature taketh no account of. There hath beene many ridiculous tales brought up of this plant, whether of old wives, or some runnagate Surgeons or Physicke-mongers I know not, (a title bad enough for them) but sure some one or moe that sought to make themselves famous and skilfull above others, were the first brochers of that errour I speake of. They adde further, That it is never or very seldome to be found growing naturally but under a gallowes, where the matter that hath fallen from the dead body hath given it the shape of a man; and the matter of a woman, the substance of a female plant; with many other such doltish dreames. They fable further and affirme, That he who would take up a plant thereof must tie a dog therunto to pull it up, which will give a great shreeke at the digging up; otherwise if a man should do it, he should surely die in short space after. Besides many fables of loving matters, too full of scurrilitie to set forth in print, which I forbeare to speake of.

All which dreames and old wives tales you shall from henceforth cast out of your bookes and memory; knowing this, that they are all and everie part of them false and most untrue: for I my selfe and my servants also have digged up, planted, and replanted very many, and yet never could either perceive shape of man or woman, but sometimes one streight root, sometimes two, and often six or seven branches comming from the maine great root, even as Nature list to bestow upon it, as to other plants. But the idle drones that have little or nothing to do but eate and drinke, have bestowed some of their time in carving the roots of Brionie, forming them to the shape of men and women: which falsifying practise hath confirmed the errour amongst the simple and unlearned people, who have taken them upon their report to be the true Mandrakes.

The female Mandrake is like unto the male, saving that the leaves hereof be of a more swart or darke greene colour: and the fruit is long like a peare, and the other like an apple.

Mandrake groweth in hot Regions, in woods and mountaines, as in mount Garganus in Apulia, and such like places; we have them onely planted in gardens, and are not elsewhere to be found in England.

Mandrake is called *Circæa*, of *Circe* the witch, who by art could procure love: for it hath beene thought that the Root hereof serveth to win love.

The wine wherein the root hath been boyled or infused provoketh sleepe and asswageth paine. The smell of the Apples moveth to sleepe likewise; but the juice worketh more effectually if you take it in small quantitie.

Great and strange effects are supposed to bee in Mandrakes, to cause women to be fruitfull and beare children, if they shall but carry the same neere to their bodies. Some do from hence ground it, for that *Rahel* desired to have her sisters Mandrakes (as the text is translated) but if we look well into the circumstances which there we shall finde, we may rather deem it otherwise. Yong *Ruben* brought home amiable and sweet-smelling floures, (for so signifieth the Hebrew word, used *Cantic*. 7. 13. in the same sence) rather for their beauty and smell, than for their vertue. Now in the floures of Mandrake there is no such delectable or amiable smell as was in these amiable floures which *Ruben* brought home. Besides, we reade not that *Rahel* conceived hereupon, for *Leah Jacobs* wife had foure children before God granted that blessing of fruitfulnesse unto *Rahel*. And last of all, (which is my chiefest

reason) *Jacob* was angry with *Rahel* when shee said, Give
mee children or els I die; and demanded of her, whether
he were in the stead of God or no, who had withheld from
her the fruit of her body. And we know the Prophet *David*
saith, Children & the fruit of the womb are the inheritance
that commeth of the Lord, *Psal.* 127.

GINGER

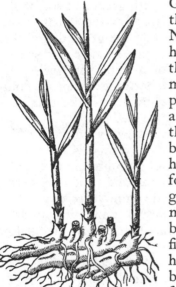

The true figure of Ginger

Ginger is most impatient of
the coldnesse of these our
Northerne regions, as my selfe
have found by proofe, for that
there have beene brought unto
me at severall times sundry
plants thereof, fresh, greene,
and full of juice, as well from
the West Indies, as from Barbary and other places; which
have sprouted and budded
forth greene leaves in my
garden in the heate of Summer, but as soone as it hath
beene but touched with the
first sharpe blast of Winter, it
hath presently perished both
blade and root. The true
forme or picture hath not before this time been set forth by
any that hath written, but the world hath beene deceived
by a counterfeit figure.

Ginger groweth in Spaine, Barbary, in the Canarie
Islands, and the Azores. Our men who sacked Domingo
in the Indies, digged it up there in sundry places wilde.
Ginger flourisheth in the hot time of Sommer, and loseth
his leaves in Winter.

197

Ginger, as *Dioscorides* reporteth, is right good with meat in sauces, or otherwise in conditures; for it is of an heating and digesting qualitie, and is profitable for the stomacke, and effectually opposeth it selfe against all darknesse of the sight; answering the qualities and effects of pepper.

H E N B A N E

The common blacke Henbane hath great and soft stalkes, leaves very broad, soft, and woolly, somewhat jagged, especially those that grow neere to the ground, and those that grow upon the stalke, narrower, smaller, and sharper, the floures are bell-fashion, of a feint yellowish white, and browne within towards the bottome: when the floures are gone, there come hard knobby husks like small cups or boxes, wherein are small brown seeds.

Blacke Henbane grows almost every where by highways, in the borders of fields about dunghils and untoiled places: the white Henbane is not found but in the gardens of those that love physicall plants: the which groweth in my garden, and doth sow it selfe from yeare to yeare.

Henbane causeth drowsinesse, and mitigateth all kinde of paine: it is good against hot & sharp distillations of the eyes and other parts.

The leaves stamped with the ointment *Populeon*, made of Poplar buds, asswageth the pain of the gout.

To wash the feet in the decoction of Henbane causeth sleepe; and also the often smelling to the floures.

The leaves, seed, and juice taken inwardly cause an unquiet sleep like unto the sleepe of drunkennesse, which continueth long, and is deadly to the party.

The root boiled with vinegre, & the same holden hot in the mouth, easeth the pain of the teeth. The seed is used by Mountibank tooth-drawers which run about the country, to cause worms come forth of the teeth, by burning it in a chafing dish of coles, the party holding his mouth over the fume thereof: but some crafty companions to gain mony convey small lute-strings into the water, persuading the patient, that those small creepers came out of his mouth or other parts which he intended to ease.

YELLOW HENBANE, OR ENGLISH TABACO

Yellow Henbane groweth to the height of two cubits: the stalke is thicke, fat, and green of colour, ful of a spongeous pith, and is divided into sundry branches, set with smooth and even leaves, thicke and full of juice. The floures grow at the tops of the branches, orderly placed, of a pale yellow color, somthing lesser than those of the black Henbane. The cups wherein the floures do stand, are like, but lesser, tenderer, and without sharpe points, wherein is set the huske or cod somwhat round, full of very smal seed like the seed of marjerom. The root is small and threddy.

Yellow Henbane is sowne in gardens, where it doth prosper exceedingly, insomuch that it cannot be destroied where it hath once sown it self, & it is dispersed into most parts of London.

It floureth in the summer moneths, and oftentimes till Autumne be farre spent, in which time the seed commeth to perfection.

Yellow Henbane is called *Hyoscyamus luteus*: of some, *Nicotiana*, of *Nicot* a Frenchman that brought the seeds from the Indies, as also the seeds of the true Tabaco, whereof this hath bin taken for a kind; insomuch that *Lobel* hath called it *Dubius Hyoscyamus*, or doubtfull Henbane, as a plant participating of Henbane and Tabaco: and it is used of divers in stead of Tabaco, and called by the same name, for that it hath bin brought from Trinidada, a place so called in the Indies, as also from Virginia and other places, for Tabaco; and doubtlesse, taken in smoke it worketh the same kind of drunkennesse that the right Tabaco doth.

This herb availeth against all botches, and such like, beeing made into an unguent or salve as followeth: Take of the greene leaves three pounds and an halfe, stampe them very smal in a stone mortar; of oile Olive one quart: set them to boile in a brasse pan or such like, upon a gentle fire, continually stirring it untill the herbs seem blacke, and wil not boile or bubble any more: then shall you have an excellent green oile; which beeing strained from the feces or drosse, put the cleare and strained oile to the fire again, adding therto of wax half a pound, of rosen foure ounces, and of good turpentine two ounces: melt them all together, and keepe it in pots for your use, to cure all cuts or hurts in the head; wherewith I have gotten both crownes and credit.

It is used of some in stead of Tabaco, but to small purpose or profit, although it doth stupifie or dull the sences, and cause that kind of giddines that Tabaco doth, and likewise spitting, which any other herb of hot temperature will do, as Rosemary, Time, Winter-Savorie, sweet Marjerome, and such like: any of the which I like better to be taken in smoke, than this kind of doubtful Henbane.

TABACO, OR HENBANE OF PERU

There be two sorts or kinds of Tabaco, one greater, the other lesser; the greater was brought into Europe out of the provinces of America, which we call the West Indies; the other from Trinidada, an Island neere unto the continent of the same Indies. Some have added a third sort, and others make the yellow Henbane a kind thereof.

Tabaco, or Henbane of Peru hath very great stalkes of the bignesse of a childes arme, growing in fertile and well dunged ground of seven or eight foot high, dividing it selfe into sundry branches of great length; whereon are placed in most comly order very faire long leaves, broad, smooth, and sharp pointed, soft, and of a light green colour, so fastned about the stalke, that they seeme to embrace and compasse it about. The floures grow at the top of the stalks, in shape like a bell-floure, somewhat long and cornered, hollow within, of a light carnation colour, tending to whitenesse toward the brims. The seed is con-tained in long sharpe pointed cods or seed-vessels like unto the seed of yellow Henbane, but somewhat smaller, and browner of colour. The root is great, thicke, and of a wooddy substance, with some threddy strings annexed there-unto.

Tabaco, or Henbane of Peru

Trinidada Tabaco hath a thicke tough and fibrous root, from which immediately rise up long broad leaves and smooth, of a greenish colour, lesse than those of Peru: among which rises up a stalk dividing it self at the ground into divers branches, wheron are set confusedly the like leaves but lesser. At the top of the stalks stand up long necked hollow floures of a pale purple tending to a blush colour: after which succeed the cods or seed-vessels, including many small seeds like unto the seed of Marjerom. The whole plant perisheth at the first approch of winter.

These were first brought into Europe out of America, which is called the West Indies, in which is the province or countrey of Peru: but being now planted in the gardens of Europe it prospers very well, and comes from seed in one yeare to beare both floures and seed. The which I take to be better for the constitution of our bodies, than that which is brought from India; & that growing in India better for the people of the same country: notwithstanding it is not so thought of our Tabaconists; for according to the English proverb, Far fetcht & dear bought is best for Ladies.

Tabaco must be sowne in the most fruitfull ground that may be found, carelesly cast abroad in sowing, without raking it into the ground, or any such pain or industry taken as is requisit in the sowing of other seeds, as my self have found by proof, who have experimented every way to cause it quickly to grow: for I have committed some to the earth in the end of March, some in Aprill, and some in the beginning of May, because I durst not hasard all my seed at one time, lest some unkindely blast should happen after the sowing, which might be a great enemie thereunto.

The people of America call it *Petun*. Some, as *Lobel* and *Pena*, have given it these Latine names, *Sacra herba*, *Sancta herba*, and *Sanasancta Indorum*. Others, as *Dodonæus*, call it *Hyoscyamus Peruvianus*, or Henbane of Peru. *Nicolaus Monardus* names it *Tabacum*. That it is *Hyoscyami species*, or a kinde of Henbane, not only the forme being like to yellow Henbane, but the qualitie also doth declare; for it bringeth drowsinesse, troubleth the sences, and maketh a man as it were drunke by taking the fume only; as *Andrew Theuet* testifieth, and common experience sheweth: of some it is called *Nicotiana*, the which I refer to the yellow Henbane for distinctions sake.

Tabaco is a remedy for the tooth-ache, if the teeth and gumbs be rubbed with a linnen cloth dipped in the juice, and afterward a round ball of the leaves laid unto the place.

The weight of foure ounces of the juice hereof drunke

procureth afterward a long and sound sleepe, as wee have learned of a friend by observation, who affirmed, That a strong countreyman of a middle age having a dropsie, took it, and being wakened out of his sleepe called for meat and drinke, and after that became perfectly cured.

Moreover, the same man reported, That he had cured many countreymen of agues, with the distilled water of the leaves drunke a little while before the fit.

Likewise there is an oile to be taken out of the leaves that healeth merri-galls, kibed heeles, and such like.

The dry leaves are used to be taken in a pipe set on fire and suckt into the stomacke, and thrust forth againe at the nosthrils, against the paines in the head, rheumes, aches in any part of the bodie, whereof soever the originall proceed, whether from France, Italy, Spaine, Indies, or from our familiar and best knowne diseases. Those leaves do palliate or ease for a time, but never perform any cure absolutely: for although they empty the body of humors, yet the cause of the griefe cannot be so taken away. But some have learned this principle, That repletion doth require evacuation; that is to say, That fulnesse craveth emptinesse; and by evacuation doe assure themselves of health. But this doth not take away so much with it this day, but the next bringeth with it more. As for example, a Well doth never yeeld such store of water as when it is most drawn and emptied. My selfe speake by proofe; who have cured of that infectious disease a great many, divers of which had covered or kept under the sickenesse by the helpe of Tabaco as they thought, yet in the end have bin constrained to have unto such an hard knot, a crabbed wedge, or else had utterly perished.

Some use to drink it (as it is termed) for wantonnesse, or rather custome, and cannot forbeare it, no not in the midst of their dinner; which kinde of taking is unwholsome and very dangerous: although to take it seldom, and that physically, is to be tolerated, and may do some good:

but I commend the syrrup above this fume or smoky medicine.

It is taken of some physically in a pipe once in a day at the most, and that in the morning fasting, against paines in the head, stomack, and griefe in the brest and lungs: against catarrhs and rheums, and such as have gotten cold and hoarsenesse.

They that have seene the proofe hereof, have credibly reported, That when the Moores and Indians have fainted either for want of food or rest, this hath bin a present remedie unto them, to supply the one, and to help them to the other.

The priests and Inchanters of the hot countries do take the fume thereof until they be drunke, that after they have lien for dead three or foure houres, they may tell the people what wonders, visions, or illusions they have seen, and so give them a prophetical direction or foretelling (if we may trust the Divell) of the successe of their businesse.

The juyce or distilled water of the first kind is very good against catarrhs, the dizzinesse of the head, and rheums that fall downe the eies, against the pain called the megram, if either you apply it unto the temples, or take one or two green leaves, or a dry leafe moistned in wine, and dried cunningly upon the embers, and laid thereto.

It cleeres the sight, and taketh away the webs and spots thereof, being annointed with the juyce bloud-warme.

The oile or juyce dropped into the eares is good against deafnesse; a cloth dipped in the same and layd upon the face, taketh away the lentils, rednesse, and spots thereof.

Many notable medicines are made hereof against the old and inveterat cough, against asthmaticall or pectorall griefes, all which if I should set downe at large, would require a peculiar volume.

It is also given unto such as are accustomed to swoune.

It is used in outward medicines, either the herbe boiled

with oile, wax, rosin, and turpentine, as before is set downe in yellow Henbane, or the extraction thereof with salt, oile, balsam, the distilled water, and such like, against tumours, apostumes, old ulcers of hard curation, botches, scabbes, stinging with nettles, carbuncles, poisoned arrowes, and wounds made with gunnes or any other weapons.

It is excellent good in burnings and scaldings with fire, water, oile, lightning, or such like, boiled with hogges grease into the forme of an ointment, as I have often prooved, and found most true; adding a little of the juice of Thorne-Apple leaves, spreading it upon a cloth and so applying it.

I doe make hereof an excellent Balme to cure deep wounds and punctures made by some narrow sharpe pointed weapon. Which Balsame doth bring up the flesh from the bottome verie speedily, and also heale simple cuts in the flesh according to the first intention, that is, to glew or soder the lips of the wound together, not procuring matter or corruption to it, as is commonly seene in the healing of wounds. The Receit is this: Take Oile of Roses, Oile of S. Johns Wort, of either one pinte, the leaves of Tabaco stamped small in a stone mortar two pounds; boile them together to the consumption of the juice, straine it and put it to the fire againe, adding thereunto of Venice Turpentine two ounces, or Olibanum and Masticke of either halfe an ounce, in most fine and subtil pouder: the which you may at all times make an unguent or salve, by putting thereto wax and rosin to give unto it a stiffe body, which worketh exceeding well in malignant and virulent ulcers, as in wounds and punctures.

I send this jewell unto you women of all sorts, especially such as cure and helpe the poore and impotent of your countrey without reward. But unto the beggarly rabble of witches, charmers, and such like couseners, that regard more to get money, than to helpe for charitie, I wish these few medicines far from their understanding, and from those deceivers, whom I wish to be ignorant herein. But courteous gentlewomen, I may not for the malice that I doe beare unto such, hide any thing from you of such importance: and therefore take one more that followeth, wherewith I have done many and good cures, although of small cost; but regard it not the lesse for that cause. Take the leaves of Tabaco two pounds, Hogs grease one pound, stampe the herbe small in a stone morter, putting thereto a small cup full of red or claret wine, stirre them well together, cover the morter from filth, and so let it rest untill morning; then put it to the fire and let it boile gently, continually stirring it untill the consumption of the wine: straine it and set it to the fire againe, putting thereto the juyce of the herbe one pound, of Venice turpentine foure ounces; boile them together to the consumption of the juice, then adde therto of the roots of round *Aristolochia* or Birthworth in most fine pouder two ounces, sufficient wax to give it a body; the which keep for thy wounded poore neighbor.

The garden Mallow called Hollyhocke

The tame or garden Mallow bringeth forth broad round leaves of a whitish greene colour, rough, and greater than those of the wilde Mallow: the stalke is streight, of the height of foure or six cubits; whereon do grow upon slender foot-stalks single floures, not much unlike to the wilde Mallow, but greater, consisting only of five leaves, sometimes white or red, now and then of a deep purple colour, varying diversly as Nature list to play with it: in their places groweth up a round knop like a little cake, compact or made up of a multitude of flat seeds like little cheeses. The root is long, white, tough, easily bowed, and groweth deep in the ground.

The double Hollihocke with purple floures hath great broad leaves, confusedly indented about the edges, and likewise toothed like a saw.

Double purple Hollihocke

These Hollihockes are sowne in gardens, almost every where, and are in vaine sought elsewhere.

The second yeere after they are sowne they bring forth their floures in July and August, when the seed is ripe the stalke withereth, the root remaineth and sendeth forth new stalkes, leaves and floures, many yeares after.

The Hollihocke is called of divers, *Rosa ultra-marina,* or outlandish Rose.

Gith, or Nigella

Nigella, which is both faire and pleasant, called Damaske Nigella, is very like unto the wilde Nigella in his small cut and jagged leaves, but his stalke is longer: the flours are greater, and every floure hath five small greene leaves under him, as it were to support and beare him up: which floures being gone, there succeed and follow knops and seed like the former, but without smell or savour.

The tame are sowne in gardens: the wilde ones doe grow of themselves among corne and other graine, in divers countries beyond the seas.

Gith is called in Italian, *Nigella*: in English, Gith, and Nigella Romana, in Cambridgeshire, Bishops wort: and also *Divæ Catherinæ flos,* Saint Katharines floure.

The seed parched or dried at the fire, brought into

pouder, and wrapped in a piece of fine lawne or sarcenet, cureth all murs, catarrhes, rheumes, and the pose, drieth the braine, and restoreth the sence of smelling unto those

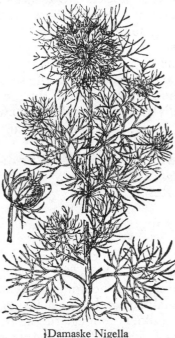

which have lost it, being often smelled unto from day to day, and made warme at the fire when it is used.

It takes away freckles, being laid on mixed with vineger. To be briefe, as *Galen* saith, it is a most excellent remedy.

It serveth well among other sweets to put into sweet waters, bagges, and odoriferous pouders.

FLOURE-GENTLE

There be divers sorts of Floure-gentle, differing in many points very notably, as in greatnesse and smalnesse; some purple, and others of a skarlet colour;

¡Damaske Nigella

and one above the rest wherewith Nature hath seemed to delight her selfe, especially in the leaves, which in variable colours strives with the Parrats feathers for beauty.

Purple Floure-gentle riseth up with a stalke a cubit high, and somtimes higher, streaked or chamfered alongst the same, often reddish toward the root, and very smooth; which divides it self toward the top into smal branches, about which stand long leaves, broad, sharpe pointed, soft, slipperie, of a greene colour, and sometimes tending to a reddish: in stead of floures come up eares or spoky tufts, very brave to look upon, but without smel, of a shining light purple, with a glosse like Velvet, but far passing it: which when they are bruised doe yeeld a juice almost of the same colour, and being gathered, doe keep their beauty a long time after; insomuch that being set in water, it will revive again as at the time of his gathering, and remaineth so, many yeares; whereupon likewise it hath taken it's name. The seed standeth in the ripe eares, of colour blacke, and much glittering: the root is short and full of strings.

Purple Floure-Gentle

It farre exceedeth my skill to describe the beauty and excellencie of this rare plant called *Floramor*; and I thinke the pensil of the most curious painter will be at a stay, when he shall come to set it downe in his lively colours. But to colour it after my best manner, this I say, *Floramor* hath a thicke knobby root, whereon do grow many threddie strings; from which riseth a thicke stalke, but tender and soft, which beginneth to divide it selfe into sundry branches at the ground and so upward, whereupon doth grow many leaves, wherein doth consist his beauty: for in few words, everie leafe resembleth in colour the most faire and beautifull feather of a Parat especially those feathers that are mixed with most sundry colours, as a stripe of red, and a line of yellow, a dash of white, and a rib of green colour, which I cannot with words set forth, such are the sundry mixtures of colours that Nature hath bestowed in her greatest jolitie, upon this floure. The floure doth grow betweene the foot-stalks of those leaves, and the body of the stalke or trunke, base, and of no moment in respect of the leaves, being as it were little chaffie husks of an overworne tawny colour: the seed is black, and shining like burnished horne.

These pleasant floures are sowne in gardens, especially for their great beautie. They floure in August, and continue flourishing till the frost overtake them, at what time they perish.

It is reported they stop all kindes of bleeding; which is not manifest by any apparant quality in them, except peradventure by the colour onely that the red eares have: for some are of opinion, that all red things stanch bleeding in any part of the body: because some things of red colour doe stop bloud: But *Galen, lib.* 2 & 4. *de simp. facult.* plainly sheweth, that there can be no certainty gathered from the colours, touching the vertues of simple and compound medicines: wherefore they are ill persuaded, that thinke the floure Gentle to stanch bleeding, because of the colour onely, if they had no other reason to induce them thereto.

Golden Rod

Golden Rod hath long broad leaves somewhat hoary and sharpe pointed; among which rise up browne stalkes two foot high, dividing themselves toward the top into sundry branches, charged or loden with small yellow floures; which when they be ripe turn into downe which is carried away with the winde.

It is extolled above all other herbes for the stopping of bloud in bleeding wounds; and hath in times past beene had in greater estimation and regard than in these daies: for in my remembrance I have known the dry herbe which came from beyond the sea sold in Bucklersbury in London for halfe a crowne an ounce. But since it was found in

Hampstead wood, even as it were at our townes end, no man will give halfe a crowne for an hundred weight of it: which plainely setteth forth our inconstancie and sudden mutabilitie, esteeming no longer of any thing, how pretious soever it be, than whilest it is strange and rare. This verifieth our English proverbe, Far fetcht and deare bought is best for Ladies. Yet it may be more truely said of phantasticall Physitions, who when they have found an approved medicine and perfect remedy neere home against any disease; yet not content therewith, they will seeke for a new farther off, and by that meanes many times hurt more than they helpe. Thus much I have spoken to bring these new fangled fellowes backe againe to esteeme better of this admirable plant than they have done, which no doubt have the same vertue now that then it had, although it growes so neere our owne homes in never so great quantity.

Marigolds

The greatest double Marigold hath many large, fat, broad leaves, springing immediatly from a fibrous or threddy root: the upper sides of the leaves are of a deepe greene, and the lower side of a more light and shining greene: among which rise up stalkes somewhat hairie, and also somewhat joynted, and full of a spungeous pith. The floures in the top are beautifull, round, very large and double, something sweet, with a certaine strong smell, of a light saffron colour, or like pure gold: from the which follow a number of long crooked seeds, especially the outmost, or those that stand about the edges of the floure; which being sowne commonly bring forth single floures, whereas contrariwise those seeds in the middle are lesser, and for the most part bring forth such floures as that was from whence it was taken.

This fruitfull or much bearing Marigold is likewise

called of the vulgar sort of women, Jacke-an-apes on horse backe: it hath leaves, stalkes, and roots like the common sort of Marigold, differing in the shape of his flours, for this plant doth bring forth at the top of the stalke one floure like the other Marigolds; from the which start forth sundry other small floures, yellow likewise, and

of the same fashion as the first, which if I be not deceived commeth to passe *per accidens*, or by chance, as Nature oftentimes liketh to play with other floures, or as children are borne with two thumbes on one hand, and such like, which living to be men, do get children like unto others; even so is the seed of this Marigold, which if it be sowen it brings forth not one floure in a thousand like the plan from whence it was taken.

The Marigold floureth from Aprill or May even untill Winter, and in Winter also, if it bee warme. It is called *Calendula*: it is to be

The great double Marigold

seene in floure in the Calends almost of every moneth: it is also called *Chrysanthemum*, of his golden colour.

The yellow leaves of the floures are dried and kept throughout Dutchland against Winter, to put into broths, in Physicall potions, and for divers other purposes, in such quantity, that in some Grocers or Spice-sellers houses are to be found barrels filled with them, and retailed by the penny more or lesse, insomuch that no broths are well made without dried Marigolds.

AFRICAN MARIGOLD

The common Africane, or as they vulgarly terme it French Marigold, hath small weake and tender branches trailing upon the ground, reeling and leaning this way and that way, beset with leaves consisting of many particular leaves, indented about the edges, which being held up against the sunne, or to the light, are seene to be full of holes like a sieve, even as those of Saint Johns woort: the floures stand at the top of the springy branches forth of long cups or husks, consisting of eight or ten small leaves, yellow underneath, on the upper side of a deeper yellow tending to the colour of a darke crimson velvet, as also soft in handling: but to describe the colour in words, it is not possible, but this way; lay upon paper with a pensill a yellow colour called Masticot, which being dry, lay the same over with a little saffron steeped in water or wine, which setteth forth most lively the colour. The whole plant is of a most ranke and unwholesome smell, and perisheth at the first frost.

They are cherished and sowne in gardens every yere: they grow every where almost in Africke of themselves, from whence wee first had them, and that was when *Charles* the fifth, Emperor of Rome made a famous conquest of Tunis.

The unpleasant smel, especially of that common sort with single floures doth shew that it is of a poisonsome and cooling qualitie; and also the same is manifested by divers experiments: for I remember, saith *Dodonæus*, that I did see a boy whose lippes and mouth when hee began to chew the floures did swell extreamely; as it hath often happened unto them, that playing or piping with quils or kexes of Hemlockes, do hold them a while betweene their lippes: likewise he saith, we gave to a cat the floures with their cups, tempered with fresh cheese, shee forthwith mightily swelled, and a little while after died: also mice that have eaten of the seed thereof have been found dead. All which things doe declare that this herbe is of a venomous and poysonsome facultie; and that they are not to be hearkned unto, that suppose this herb to be a harmlesse plant: so to conclude, these plants are most venomous and full of poison, and therefore not to be touched or smelled unto, much lesse used in meat or medicine.

FLOURE OF THE SUN, OR THE MARIGOLD OF PERU

The greater Sun-floure

The Indian Sun, or the golden floure of Peru, is a plant of such stature and talnesse, that in one summer, beeing sowne of a seed in Aprill, it hath risen up to the height of fourteene foot in my garden, where one floure was in weight three pound and two ounces, & crosse overthwart the floure by measure sixteen inches broad. The stalks are upright & straight, of the bignesse of a strong mans arme, beset with large leaves even to the top, like unto the great Clot bur: at the top of the stalk commeth forth for the most part one floure, yet many times there spring out sucking buds which come to no perfection: this great floure is in shape like to the Camomil floure, beset round about with a pale or border of goodly yellow leaves, in shape like the leaves of the floures of white

Lillies: the middle part whereof is made as it were of unshorn velvet, or some curious cloath wrought with the needle: which brave worke, if you do thorowly view and marke well, it seemeth to be an innumerable sort of small floures, resembling the nose or nosle of a candlestick broken from the foot thereof; from which small nosle sweats forth excellent fine and cleare turpentine, in sight, substance, savor, and tast. The whole plant in like manner being broken smelleth of turpentine: when the plant groweth to maturitie the floures fall away, in place whereof appeareth the seed, black and large, much like the seed of Gourds, set as though a cunning workman had of purpose placed them in very good order, much like the honycombs of Bees.

These plants grow of themselves without setting or sowing, in Peru, and in divers other provinces of America, from whence the seeds have beene brought into these parts of Europ. There hath bin seen in Spain and other hot regions a plant sowne and nourished up from seed, to attaine to the height of 24 foot in one yeare. The seed must be set or sowne in the beginning of April, if the weather be temperat, in the most fertill ground that may be, and where the Sun hath most power the whole day.

The flour of the Sun is called in Latine *Flos Solis*, for that some have reported it to turn with the Sun, which I could never observe, although I have indeavored to finde out the truth of it: but I rather thinke it was so called because it resembles the radiant beams of the Sunne, whereupon some have called it *Corona Solis*, and *Sol Indianus*, the Indian Sunne-floure: others, *Chrysanthemum Peruvianum*, or the Golden floure of Peru: in English, the floure of the Sun, or the Sun-floure.

There hath not any thing bin set down either of the antient or later writers, concerning the vertues of these plants, notwithstanding we have found by triall, that the buds before they be floured boiled and eaten with butter, vineger, and pepper, after the manner of Artichokes, are exceeding pleasant meat.

BLEW-BOTTLE OR CORNE-FLOURE

The great Blew-Bottle hath long leaves smooth, soft, downy, and sharp pointed: among the leaves rise up crooked and pretty thicke branches, chamfered, furrowed, and garnished with such leaves as are next the ground: on the tops wherof stand faire blew flours tending to purple, consisting of divers little flours, set in a scaly huske or knap like those of Knapweed: the seed is rough or bearded at one end, smooth at the other.

The common Corn-floure hath leaves somwhat hackt or cut in the edges: the floures grow at the top of the stalks, of a blew colour: the seed is smooth, bright shining, and wrapped in a woolly or flocky matter.

The first groweth in my garden, and in the gardens of Herbarists, but not wilde that I know of. The other grows in corn fields among Wheat, Rie, Barley, and other graine: it is sowne in gardens, and by cunning looking to doth oft times become of other colours, and some also double.

The old Herbarists call it *Cyanus flos*, of the blew colour which it naturally hath: in Italian, *Baptisecula*, as though it should be called *Blaptisecula*, because it hindereth and annoyeth the Reapers, by dulling and turning the edges of their sicles in reaping of corne: in English it is called Blew-Bottle, Blew-Blow, Cornefloure, and hurt-Sicle.

CORNE

This kinde of Wheate is the most principall of all other, whose eares are altogether bare or naked, without awnes or chaffie beards. The stalke riseth from a threddy root, compact of many strings, joynted or kneed at sundry distances; from whence shoot forth grassie blades and leaves like unto Rie, but broader. The plant is so well knowne to many, and so profitable to all, that the meanest and most ignorant need no larger description to know the same by.

Wheat (saith *Galen*) is very much used of men, and with greatest profit. Those Wheats do nourish most that be hard, and have their whole sub-stance so closely compact as they can scarsely be bit asunder; for such do nourish very much: and the contrary but little.

Slices of fine white bread laid to infuse or steepe in Rose water, and so applied unto sore eyes which have many hot humours falling into them, doe easily defend the humour, and cease the paine.

The oyle of wheat pressed forth betweene two plates of hot iron, healeth the chaps and chinks of the hands,

Bright Wheat

feet, and fundament, which come of cold, making smooth the hands, face or any other part of the body.

OTES

Avena Vesca, common Otes, is called *Vesca, a Vescendo*, because it is used in many countries to make sundry sorts of bread, as in Lancashire, where it is their chiefest bread corne for Jannocks, Haver cakes, Tharffe cakes, and those which are called generally Oten cakes; and for the most part they call the graine Haver, whereof they do likewise make drinke for want of Barley.

Avena Nuda is like unto the common Otes; differing in that, that these naked Otes immediately as they be threshed, without helpe of a Mill become Otemeale fit for our use. In consideration whereof in Northfolke and Southfolke they are called unhulled or naked Otes. Some of those good house-wives that delight not to have any thing but from hand to mouth, according to our English proverbe, may (while their pot doth seeth) go to the barne, and rub forth with their hands sufficient for that present time, not willing to provide for to morrow, according as the scripture speaketh, but let the next day bring it forth.

Common Otes put into a linnen bag, with a little bay salt quilted handsomely for the same purpose, and made hot in a frying pan, and applied very hot, easeth the paine in the side called the stitch.

Otemeale is good for to make a faire and wel coloured maid to looke like a cake of tallow, especially if she take next her stomacke a good draught of strong vinegre after it.

CANARIE SEED, OR PETY PANICK

Canarie seed, or Canarie grasse after some, hath many small hairy roots, from which arise small strawy stalkes joynted like corne, whereupon doe grow leaves like those of Barley, which the whole plant doth very well resemble. The small chaffie eare groweth at the top of the stalk, wherein is contained small seeds like those of Panick, of a yellowish colour and shining.

Shakers or Quaking Grasse groweth to the height of halfe a foot, and sometimes higher, when it groweth in fertile medowes. The stalke is very small and benty, set with many grassie leaves like the common medow-grasse, bearing at the top a bush or tuft of flat scaly pouches, like those of Shepheards purse, but thicker, of a browne colour, set upon the most small and weak hairy footstalks that may be found, whereupon those small pouches do hang; by means of which small hairy strings, the knaps which are the floures do continually tremble and shake, in such sort that it is not possible with the most stedfast hand to hold it from shaking.

Canary seed groweth naturally in Spain, and also in the Fortunat or Canary Islands, and also in England or any other of these cold regions, if it be sowne therein. Quaking *Phalaris* groweth in fertile pasture, and in dry medowes.

These Canarie seeds are sowne in May, and are ripe in August.

Canarie seed or Canarie Corne is called of the Latines *Phalaris*: in English, Canarie seed, and Canarie Grasse. *Phalaris pratensis*

Quaking Grasse

is called also *Gramen tremulum*: in Cheshire about Nantwich, Quakers and Shakers: in some places, Cow-Quakes.

Apothecaries, for want of Millet, do use Canary Seed with good successe in fomentations; for in dry fomentations it serveth in stead thereof, and is his *succedaneum*, or *quid pro quo*. We use it in England also to feed Canary Birds.

HEATH, HATHER, OR LINGE

Small leafed Heath

There be divers sorts of Heath, some greater, some lesser; some with broad leaves, and some narrower; some bringing forth berries, and others nothing but floures.

The common Heath is a low plant, but yet wooddy and shrubby, scarce a cubit high: it brings forth many branches, whereupon doe grow sundry little leaves somewhat hard and rough, very like to those of Tamariske or the Cypres tree: the floures are orderly placed alongst the branches, small, soft, and of a light red colour tending to purple: the root is also wooddy, and creepeth under the upper crust of the earth: and this is the Heath which the Antients tooke to be the right and true Heath.

There is another Heath which differeth not from the precedent, saving that this plant bringeth forth floures as white as snow, wherein consisteth the difference: wherefore we may cal it *Erica pumila alba*, Dwarfe Heath with white floures.

Crossed Heath growes to the height of a cubit and a halfe, full of branches, commonly lying along upon the ground, of a dark swart colour: whereon grow small leaves set at certain spaces two upon one side, and two on the other, opposite, one answering another, even as do the leaves of Crossewort. The floures in like manner stand along the branches crosse-fashion, of a dark overworne greenish colour. The root is likewise wooddy, as is all the rest of the plant.

The steeple Heath hath likewise many wooddy branches garnished with small leaves that easily fall off from the dried stalks; among which come forth divers little mossie greenish floures of small moment. The whole bush for the most part groweth round together like a little cock of hay, broad at the lower part and sharp above like a pyramide or steeple, whereof it tooke his name.

The small or thinne leafed Heath is also a low and base shrub, having many small and slender shoots comming from the root, of a reddish browne colour; whereupon doe grow very many small leaves not unlike to them of common Tyme, but much smaller and tenderer: the floures grow in tufts at certaine spaces, of a purple colour.

Heath groweth upon dry mountaines which are hungry and barren, as upon Hampsteed Heath neere London, where all the sorts do grow, except that with the white floures, and that which beareth berries. Heath with the white floures groweth upon the downes neere unto Gravesend.

These kindes or sorts of Heath do for the most part floure all the Summer, even untill the last of September.

The tender tops and floures, saith *Dioscorides*, are good to be laid upon the bitings and stingings of any venomous beast: of these floures the Bees do gather bad hony.

DWARFE KINDES OF CISTUS

The English dwarfe Cistus is a low and base plant creeping upon the ground, having many small tough branches of a browne colour; wherupon grow little leaves set together by couples, thicke, fat, and ful of substance, and covered over with a soft downe; from the bosome whereof come forth other lesser leaves: the floures before they be open are small knops or buttons, of a browne colour mixed with yellow, and beeing open and spred abroad are like those of the wilde Tansie, & of a yellow colour, with some yellower chives in the middle.

Valerius Cordus nameth it *Helianthemum*, and *Solis flos* or Sun-floure. *Pliny* writeth, that *Helianthemum* growes in the champian country Temiscyra in Pontus, and in

The white dwarfe Cistus of Germanie

the mountains of Cilicia neere the sea: saying further, that the wise men of those countries & the Kings of Persia do anoint their bodies herewith, boiled with Lions fat, a little Saffron, and Wine of Dates, that they may seem faire and beautifull; and therefore have they called it *Heliocaliden*, or the beauty of the Sun.

CLOWNES WOUND-WORT, OR ALL-HEALE

Clownes All-heale, or the Husbandmans Wound-wort, hath long slender square stalkes of the height of two cubits: at the top of the stalkes grow the floures spike fashion, of a purple colour mixed with some few spots of white, in forme like to little hoods.

It groweth in moist medowes by the sides of ditches, and likewise in fertile fields that are somewhat moist, almost every where; especially in Kent about South-fleet, neer to Gravesend, and likewise in the medowes by Lambeth neere London. It floureth in August, and bringeth his seed to perfection in the end of September.

The leaves hereof stamped with *Axungia* or hogs grease, and applied unto greene wounds in manner of a pultesse, heale them in short time, and in such absolute manner, that it is hard for any that have not had the· experience thereof to beleeve: for being in Kent about a Patient, it chanced that a poore man in mowing of Peason did cut his leg with a sithe, wherein hee made a wound to the bones, and withall very large and wide, and also with great effusion of bloud; the poore man crept unto this herbe, which he bruised with his hands, and tied a great quantitie of it unto the wound with a piece of his shirt, which presently stanched the bleeding, and ceased the paine, insomuch that the poore man presently went to his daies worke againe, and so did from day to day, without resting one day untill he was perfectly whole; which was accomplished in a few daies, by this herbe stamped with a little hogs grease, and so laid upon it in manner of a pultesse, which did as it were glew or sodder the lips of the wound together, and heale it according to the first intention, as wee terme it, that is, without drawing or bringing the wound to suppuration or matter; which was fully performed in seven daies, that would have required forty daies with balsam it selfe. I saw the wound and offered to heale the same for charity; which he refused, saying that I could not heale it so well as himselfe: a clownish answer I confesse, without any thankes for my goodwill: whereupon I have named it Clownes Wound-wort, as aforesaid. Since which time my selfe have cured many grievous wounds, and some mortall, with the same herbe; one for example done upon a Gentleman of Grayes Inne in Holborne, Mr. *Edmund Cartwright*, who was thrust into the lungs, the wound entring in at the lower part of the *Thorax*, or the brest-blade, even through that cartilaginous substance called *Mucronata Cartilago*, insomuch that from day to day the frothing and puffing of the lungs did spew forth of the wound such excrements as it was possessed of, besides the Gentleman was most dangerously vexed with a double quotidian fever; whom by Gods permission I perfectly cured in very short time, and with this Clownes experiment, and some of my foreknowne helpes, which were as followeth.

First I framed a slight unguent hereof thus: I tooke foure handfulls of the herbe stamped, and put them into a pan, whereunto I added foure ounces of Barrowes grease, halfe a pinte of oyle Olive, wax three ounces, which I boyled unto the consumption of the juyce (which is knowne when the stuffe doth not bubble at all) then did I straine it, putting it to the fire againe, adding thereto two ounces of Turpentine, the which I suffered to boile a little, reserving the same for my use.

The which I warmed in a sawcer, dipping therein small soft tents, which I put into the wound, defending the parts adjoyning with a plaister of *Calcitheos*, relented with oyle of roses: which manner of dressing and preserving I used even untill the wound was perfectly whole: notwithstanding once in a day I gave him two spoonfulls of this decoction following.

I tooke a quart of good Claret wine, wherein I boyled an handfull of the leaves of *Solidago Saracenica*, or Saracens consound, and foure ounces of honey, whereof I gave him in the morning two Spoonefulls to drinke in a small draught of wine tempered with a little sugar.

In like manner I cured a Shoo-makers servant in Holborne, who intended to destroy himselfe for causes knowne unto many now living: but I deemed it better to

cover the fault, than to put the same in print, which
might move such a gracelesse fellow to attempt the like:
his attempt was thus; First, he gave himselfe a most
mortall wound in the throat, in such sort, that when I
gave him drinke it came forth at the wound, which likewise did blow out the candle: another deepe and grievous
wound in the brest with the said dagger, and also two
others in *Abdomine*: the which mortall wounds, by Gods
permission, and the vertues of this herbe, I perfectly
cured within twenty daies: for the which the name of
God be praised.

PIMPERNELL

Pimpernell

Pimpernell is like
unto Chickweed; the
stalkes are foure
square, trailing here
and there upon the
ground, whereupon
do grow broad
leaves, and sharpe
pointed set together
by couples: from
the bosomes whereof
come forth slender
tendrels whereupon
doe grow small pur-
ple floures tending
to rednesse: which
being past there suc-
ceed fine round bul-
lets, like unto the
seed of Coriander, wherein is contained small dusty seed.
The root consisteth of slender strings.

The female Pimpernell differeth not from the male in
any one point, but in the colour of the floures; for like

225

as the former hath reddish floures, this plant bringeth forth floures of a most perfect blew colour; wherein is the difference.

They grow in plowed fields neere path waies, in Gardens and Vineyards almost every where.

They floure in Summer, and especially in the moneth of August, at what time the husbandmen having occasion to go unto their harvest worke, will first behold the floures of Pimpernell, whereby they know the weather that shall follow the next day after; as for example, if the floures be shut close up, it betokeneth raine and foule weather; contrariwise, if they be spread abroad, faire weather.

Both the sorts of Pimpernell are of a drying faculty without biting, and somewhat hot, with a certaine drawing quality, insomuch that it doth draw forth splinters and things fixed in the flesh.

The juyce cures the toothach being snift up into the nosethrils, especially into the contrary nosethrill.

DIVELS BIT

Divels bit hath small upright round stalkes of a cubite high, beset with long leaves somwhat broad, very little or nothing snipt about the edges, somwhat hairie and even. The floures also are of a dark

Divels bit

purple colour, fashioned like the floures of Scabious: the seeds are smal and downy, which being ripe are carried away with the winde.

The root is blacke, thick, hard and short, with many threddie strings fastned thereto. The great part of the root seemeth to be bitten away: old fantasticke charmers report, that the divel did bite it for envie, because it is an herbe that hath so many good vertues, and is so beneficial to mankinde.

Divels bit groweth in dry medows and woods, & about waies sides. It floureth in August, and is hard to be knowne from Scabious, saving when it floureth.

It is commonly called *Morsus Diaboli*, or Divels bit, of the root (as it seems) that is bitten off: for the superstitious people hold opinion, that the divell for envy that he beareth to mankinde, bit it off, because it would be otherwise good for many uses.

E YE-BRIGHT

Euphrasia or Eye-bright is a small low herbe not above two handfulls high, full of branches, covered with little blackish leaves dented or snipt about the edges like a Saw. The floures are small and white, sprinkled and poudered on the inner side, with yellow and purple speckes mixed therewith. The root is small and hairie.

This plant groweth in dry medowes, in greene and grassie waies and pastures standing against the Sun. Eye-bright beginneth to floure in August, and continueth unto September, and must bee gathered while it floureth for physicks use.

It is very much commended for the eies. Being taken it selfe alone, or any way else, it preserves the sight, and being feeble & lost it restores the same: it is given most fitly being beaten into pouder; oftentimes a like quantitie of Fennell seed is added thereto, and a little mace, to the which is put so much sugar as the weight of them all commeth to.

Eye-bright stamped and laid upon the eyes, or the

juice thereof mixed with white Wine, and dropped into the eyes, or the distilled water, taketh away the darknesse and dimnesse of the eyes, and cleareth the sight.

Three parts of the pouder of Eye-bright, and one part of maces mixed therewith, taketh away all hurts from the eyes, comforteth the memorie, and cleareth the sight, if halfe a spoonefull to be taken every morning fasting with a cup of white wine.

MARJEROME

Wilde Marjerome of Candy

Sweet Marjerome is a low and shrubbie plant, of a whitish colour & marvellous sweet smell, a foot or somwhat more high. The stalkes are slender, and parted into divers branches, about which grow forth little leaves soft and hoarie: the floures grow at the top in scalie or chaffie spiked eares, of a white colour. The whole plant and everie part thereof is of a most pleasant tast and aromaticall smell, and perisheth at the first approch of Winter.

These plants do grow in Spain, Italy, Candy, and other Islands thereabout, wild, and in the fields; from whence wee have the seeds for the gardens of our cold countries. They are to be watered in the middle of the day, when the Sun shineth hottest, even as Basill should be, and not in the evening nor morning, as most plants are.

Bastard Marjerome of Candy hath many threddy roots; from which rise up divers weake and feeble branches trailing upon the ground, set with faire greene leaves, not unlike those of Penny Royall, but broader and shorter: at the top of those branches stand scalie or chaffie eares of a purple colour. The whole plant is of a most pleasant sweet smell. The root endured in my garden and the leaves also greene all this Winter long, 1597. although it hath beene said that it doth perish at the first frost, as sweet Marjerome doth.

English wilde Marjerome is exceedingly well knowne to all, to have long, stiffe, and hard stalkes of two cubits high, set with leaves like those of sweet Marjerome, but broader and greater, of a russet greene colour, on the top of the branches stand tufts of purple flowers, composed of many small ones set together very closely umbell fashion.

Sweet Marjerome is a remedy against cold diseases of the braine and head, being taken any way to your best liking; put up into the nosthrils it provokes sneesing, and draweth forth much baggage flegme: it easeth the tooth-ache being chewed in the mouth.

The leaves boiled in water, and the decoction drunke, easeth such as are given to overmuch sighing. The leaves dried and mingled with honey put away black and blew markes after stripes and bruses, being applied thereto.

The leaves are excellent good to be put into all odoriferous ointments, waters, pouders, broths and meates. The dried leaves poudered, and finely searched, are good to put into Cerotes, or Cere-clothes, and ointments, profitable against cold swellings, and members out of joynt.

There is an excellent oyle to be drawne forth of these herbes, good against the shrinking of sinewes, crampes, convulsions, and all aches proceeding of a colde cause.

Bastard Marjerome is called in shops *Origanum Hispanicum*, Spanish Organy. Organy given in wine is a remedy against the bitings, and stingings of venomous beasts, and cureth them that have drunke *Opium*, or the juyce of blacke Poppy, or hemlockes, especially if it be given with wine and raisons of the sunne. It is profitably used in a looch, or a medicine to be licked, against the old cough and the stuffing of the lungs.

The juyce mixed with a little milke, being poured into the eares, mitigateth the paines thereof. The same mixed with the oile of *Ireos*, or the roots of the white Florentine floure-de-luce, and drawne up into the nosthrils, draweth downe water and flegme: the herbe strowed upon the ground driveth away serpents.

These plants are easie to be taken in potions, and therefore to good purpose they may be used and ministred unto such as cannot brooke their meate, and to such as have a sowre squamish and watery stomacke, as also against the swouning of the heart.

PENNIE ROYALL, OR PUDDING GRASSE

Pulegium regium vulgatum is so exceedingly well knowne to all our English Nation, that it needeth no description, being our common Pennie Royall.

The common Penny Royall groweth naturally wild in moist and overflown places, as in the Common neere London called Miles end, about the holes and ponds thereof in sundry places, from whence poore women bring plenty to sell in London markets; and it groweth in sundry other Commons neere London likewise.

If you have when you are at the sea Penny Royall in great quantitie dry, and cast it into corrupt water, it helpeth it much, neither will it hurt them that drinke thereof.

A Garland of Pennie Royall made and worne about the

head is of great force against the swimming in the head,
and the paines and giddinesse thereof.

VERVAINE

The stalke of upright Vervaine riseth from the root
single, cornered, a foot high, seldome above a cubit,
and afterwards divided into
many branches. The leaves
are long, greater than those
of the Oke, but with bigger
cuts and deeper: the floures
along the sprigs are little,
blew, or white, orderly
placed: the root is long, with
strings growing on it.

Creeping Vervaine send-
eth forth stalkes like unto
the former, now and then
a cubit long, cornered, more
slender, for the most part
lying upon the ground. The
leaves are like the former,
but with deeper cuts, and
more in number. The
floures at the tops of the
sprigs are blew, and purple
withall, very small as those

Vervaine

of the last described, and placed after the same manner
and order. The root groweth straight downe, being
slender and long, as is also the root of the former.

Both of them grow in untilled places neere unto
hedges, high-waies, and commonly by ditches almost
every where.

Vervaine is called in Latine, *Verbena*, and *Sacra herba*:
Verbenæ are any manner of herbes that were taken from

the Altar, or from some holy place, which because the Consull or Pretor did cut up, they were likewise called *Sagmina*, which oftentimes are mentioned in *Livy* to be grassie herbes cut up in the Capitoll. In English, Juno's teares, Mercuries moist bloud, Holy-herbe; and of some, Pigeons grasse, or Columbine, because pigeons are delighted to be amongst it, as also to eat thereof, as *Apuleius* writeth.

It is reported to be of singular force against the Tertian and Quartaine Fevers: but you must observe mother *Bombies* rules, to take just so many knots or sprigs, and no more, lest it fall out so that it do you no good, if you catch no harme by it. Many odde old wives fables are written of Vervaine tending to witchchraft and sorcery, which you may reade elsewhere, for I am not willing to trouble your eares with reporting such trifles, as honest eares abhorre to heare.

Most of the later Physitions do give the juice or decoction hereof to them that have the plague: but these men are deceived, not only in that they looke for some truth from the father of falshood and leasings, but also because in stead of a good and sure remedy they minister no remedy at all; for it is reported, that the Divell did reveale it as a secret and divine medicine.

MINTS

There be divers sorts of Mints, some of the garden, others wilde or of the field; and also some of the water.

The tame or garden Mint commeth up with stalks foure square, of an obscure red colour somewhat hairy, which are covered with round leaves nicked in the edges like a Saw, of a deep green colour: the floures are little and red, and grow about the stalkes circle-wise as those of Penny-Royall: the root creepeth aslope in the ground, having some strings on it, and now and then in sundry places it buddeth out afresh: the whole herb is of a pleasant smel, and it rather lieth downe than standeth up.

Cat-Mint or Nep growes high: the floures are of a whitish colour, set in manner of an eare or catkin: the whole herb is soft, and covered with a white down. It groweth about the borders of gardens and fields, neere to rough banks, ditches, and common wayes: it is delighted with moist and watery places, and is brought into gardens.

Water Mint is a kinde of Wilde Mint like to garden Mint: the floures in the tops of the branches are gathered together into a round eare, of a purple colour.

Mints doe floure and flourish in Summer: in winter the roots only remain: being once set, they continue long, and remaine sure and fast in the ground.

The smell of Mint, saith *Pliny,* doth stir up the minde, and the taste to a greedy desire of meat. Mint is marvellous wholesome for the stomacke. It is good against watering eies. It is poured into the eares with honied water.

It is applied with salt to the bitings of mad dogs. It will not suffer milke to cruddle in the stomacke (*Pliny* addeth, to wax soure) therefore it is put in milke that is drunke, lest those that drinke thereof should be strangled. It is laid to the stinging of wasps with good successe.

The later Herbarists doe call Nep *Herba Cattaria, &* *Herba Catti,* because cats are very much delighted herewith; for the smell of it is so pleasant unto them, that they rub themselves upon it, & wallow or tumble in it, and also feed on the branches and leaves very greedily.

It is a present helpe for them that be bursten inwardly of some fall received from an high place, and that are very much bruised, if the juice be given with wine or meade.

The savor or smell of the Water Mint rejoyceth the heart of man, for which cause they use to strew it in chambers and places of recreation, pleasure, and repose, and where feasts and banquets are made.

233

COTTON THISTLE

The common Thistle, whereof the greatest quantity of down is gathered for divers purposes, as well by the poore to stop pillowes, cushions, and beds for want of feathers, as also bought of the rich upholsters to mix with the feathers and down they do sell, which deceit would be looked unto: this Thistle hath great leaves, long and

Cotton Thistle

broad, gashed about the edges, and set with sharpe and stiffe prickles all alongst the edges, covered all over with a soft cotton or downe: out from the middest whereof riseth up a long stalke about two cubits high, cornered, and set with filmes, and also full of prickles: the heads are likewise cornered with prickles, and bring forth floures consisting of many whitish threds: the seed which succeedeth them is wrapped up in downe; it is long, of a light crimson colour, and lesser than the seed of bastard Saffron: the root groweth deep in the ground, being white, hard, wooddy, and not without strings.

These Thistles grow by high waies sides, and in ditches almost every where. They floure from June untill August, the second yeare after they be sowne: and in the mean time the seed waxeth ripe, which being thorow ripe the herbe perisheth, as doe likewise most of the other Thistles, which live no longer than till the seed be fully come to maturity.

234

This Thistle is called in English, Cotton-Thistle, white Cotton-Thistle, wilde white Thistle, Argentine or the silver Thistle.

ARTICHOKE

The Artichoke is to be planted in a fat and fruitfull soile: they doe love water and moist ground. They commit great error who cut away the side or superfluous leaves that grow by the sides, thinking thereby to increase the greatnesse of the fruit, when as in truth they deprive the root from much water by that meanes, which would nourish it to the feeding of the fruit; for if you marke the trough or hollow channell that is in every leafe, it shall appeare very evidently, that the Creator in his secret wisedome did ordaine those furrows, even from the extreme point of the leafe to the ground where it is fastned to the root, for no other purpose but to guide and leade that water which falls farre off, unto the root; knowing that without such store of water the whole plant would wither, and the fruit pine away and come to nothing.

They are planted for the most part about the Kalends of November, or somewhat sooner. The plant must bee set and dunged with good store of ashes, for that kinde of dung is thought best for planting thereof. Every yeare the slips must be torne or slipped off from the body of the root, and these are to be set in April, which will beare fruit about August following, as *Columella, Paladius,* and common experience teacheth.

The nailes, that is, the white and thicke parts which are in the botome of the outward scales or flakes of the fruit of the Artichoke, and also the middle pulpe whereon the downy seed stands, are eaten both raw with pepper and salt, and commonly boyled with the broth of fat flesh, with pepper added, and are accounted a dainty dish, being pleasant to the taste: so likewise the middle ribs of the leaves being made white and tender by good cherishing and looking to, are brought to the table as a great service together with other junkets: they are eaten with pepper and salt as be the raw Artichokes: yet both of them are of ill juyce; for the Artichoke containeth plenty of cholericke juyce, and hath an hard substance, insomuch as of this is ingendred melancholy juyce, and of that a thin and cholericke bloud, as *Galen* teacheth in his booke of the faculties of nourishments. But it is best to eate the Artichoke boyled: the ribbes of the leaves are altogether of an hard substance: they yeeld to the body a raw and melancholy juyce, and containe in them great store of winde.

SUGAR-CANE

Sugar Cane is a pleasant and profitable Reed, having long stalkes seven or eight foot high, joynted or kneed like unto the great Cane; the leaves come forth of every joynt on every side of the stalke one, like unto wings, long, narrow, and sharpe pointed. The Cane it selfe, or stalke is not hollow as the other Canes or Reeds are, but full, and stuffed with a spongeous substance in taste exceeding sweet.

The Sugar Cane groweth in many parts of Europe at this day, as in Spaine, Portugal, Olbia, and in Provence. It groweth also in Barbarie, generally almost every where in the Canarie Islands, and in those of Madera, in the East and West Indies, and many other places. My selfe did plant some shoots thereof in my garden, and some in Flanders did the like: but the coldnesse of our clymat made an end of mine, and I think the Flemmings will have the like profit of their labour.

This Cane is planted at any time of the yeare in those hot countries where it doth naturally grow, by reason they feare no frosts to hurt the young shoots at their first planting.

¶ Of the juyce of this Reed is made the most pleasant and profitable sweet, called Sugar, whereof is made infinite confections, confectures, Syrups and such like, as also preserving and conserving of sundry fruits, herbes, and floures, as Roses, Violets, Rosemary floures, and such like, which still retaine with them the name of Sugar, as Sugar Roset, Sugar Violet, &c. The which to write of would require a peculiar volume, and not pertinent unto this historie, for that it is not my purpose to make of my booke a Confectionary, a Sugar Bakers furnace, a Gentlewomans preserving pan, nor yet an Apothecaries shop or Dispensatorie; but onely to touch the chiefest matter that I purposed to handle in the beginning, that is, the nature, properties, and descriptions of plants. Notwithstanding I thinke it not amisse to shew unto you the ordering of these reeds when they be new gathered, as I received it from the mouth of an Indian my servant: he saith, They cut them in small pieces, and

Sugar Cane

put them into a trough made of one whole tree, wherein they put a great stone in manner of a mill-stone, whereunto they tie a horse, buffle, or some other beast which draweth it round: in which trough they put those pieces of Canes, and so crush and grind them as we doe the barkes of trees for Tanners, or apples for Cyder. But in some places they use a great wheele wherein slaves doe tread and walke as dogs do in turning the spit: and some others doe feed as it were the bottome of the said wheele, wherein are some sharpe or hard things which doe cut and crush the Canes into powder. And some likewise have found the invention to turne the wheele with water works, as we doe our Iron mills. The Canes being thus brought into dust or powder, they put them into great cauldrons with a little water, where they boile untill there be no more sweetnesse left in the crushed reeds. Then doe they straine them through mats or such like things, and put the liquor to boile againe unto the consistence of hony, which being cold is like unto sand both in shew and handling, but somewhat softer; and so afterwards it is carried into all parts of Europe, where it is by the Sugar Bakers artificially purged and refined to that whitenesse as we see.

Beets

The common white Beet hath great broad leaves, smooth and plain: from which rise thicke crested or chamfered stalks: the floures grow along the stalks clustering together, in shape like little stars, which being past, there succeed round & uneven prickly seed. The root is thicke, hard, and great.

There is likewise another sort hereof, that was brought unto me from beyond the seas, by that courteous Merchant master *Lete*, before remembred, the which hath leaves very great, and red of colour, as is all the rest of the plant, as well root, as stalke, and floures full of a perfect purple juyce tending to rednesse: the middle rib of which leaves are for the most part very broad and thicke, like the middle part of the Cabbage leafe, which is equall in goodnesse with the leaves of Cabbage being boyled. It grew with me 1596. to the height of eight cubits, and did bring forth his rough and uneven seed very plentifully: with which plant nature doth seeme to play and sport herselfe: for the seeds taken from that plant, which was altogether of one colour and sowen, doth bring forth plants of many and variable colours, as the worshipfull Gentleman master *John Norden* can very well testifie: unto whom I gave some of the seeds aforesaid, which in his garden brought forth many other of beautifull colours.

The Beete is sowne in gardens: it loveth to grow in a moist and fertile ground.

Being eaten when it is boyled, it nourisheth little or nothing, and is not so wholesome as Lettuce.

The juyce conveighed up into the nosthrils doth gently draw forth flegme, and purgeth the head.

The greater red Beet or Roman Beet, boyled and eaten with oyle, vinegre and pepper, is a most excellent and delicat sallad: but what might be made of the red and beautifull root (which is to be preferred before the leaves, as well in beautie as in goodnesse) I refer unto the curious and cunning cooke, who no doubt when hee had the view thereof, and is assured that it is both good and wholesome, will make thereof many and divers dishes, both faire and good.

HOPS

The Hop doth live and flourish by embracing and taking hold of poles, pearches, and other things upon which it climeth. It bringeth forth very long stalkes, rough, and hairie; also rugged leaves broad like those of the Vine, or rather of Bryony, but yet blacker, and with fewer dented divisions: the floures hang downe by clusters from the tops of the branches, puffed up, set as it were with scales like little canes, or scaled Pine apples, of a whitish colour tending to yellownesse, strong of smell: the roots are slender, and diversly folded one within another.

The Hop joyeth in a fat and fruitfull ground: also it groweth among briers and thornes about the borders of fields, I meane the wilde kinde.

The floures of hops are gathered in August and

Hops

September, and reserved to be used in beere: in the Spring time come forth new shoots or buds: in the Winter onely the roots remaine alive.

The buds or first sprouts which come forth in the Spring are used to be eaten in sallads.

The floures are used to season Beere or Ale with, and too many do cause bitternesse thereof, and are ill for the head.

The floures make bread light, and the lumpe to be sooner and easilier leavened, if the meale be tempered with liquor wherein they have been boiled.

The manifold vertues of Hops do manifestly argue the wholesomenesse of beere above ale; for the hops rather make it a physicall drinke to keepe the body in health, than an ordinary drinke for the quenching of our thirst.

ALMOND TREE

The Almond tree is like the Peach-tree, yet it is higher, bigger, of longer continuance: the leaves be very long, sharpe pointed, snipt about the edges like those of the Peach tree: the floures be alike: the fruit is also like a Peach, having on one side a cleft, with a soft skin without, and covered with a thin cotton, but under this there is none, or very little pulpe, which is hard like a gristle not eaten: the nut or stone within is longer than that of the Peach, not so rugged, but smooth; in which is contained the kernel, in taste sweet, and many times bitter; the root of the tree groweth deepe: the gum which soketh out hereof is like that of the Peachtree.

The naturall place of the Almond is in the hot regions, yet we have them in our London gardens and orchards in great plenty. The Almond floureth betimes with the Peach: the fruit is ripe in August.

Almonds taken before meate nourish but little; notwithstanding many excellent meates and medicines are therewith made for sundry griefes, yea very delicat and wholsome meates, as Almond butter, creame of Almonds, marchpane, and such like.

They doe serve to make the Physicall Barley Water, and Barley Creame, which are given in hot Fevers, as also for other sicke and feeble persons, for their further refreshing and nourishments.

The oyle which is newly pressed out of the sweet Almonds is a mitigater of paine and all maner of aches. The oile of Almonds makes smooth the hands and face of delicat persons, and clenseth the skin from all spots, pimples, and lentils.

And it is reported that five or six Almonds being taken fasting do keepe a man from being drunke. These also clense and take away spots and blemishes in the face, and in other parts of the body.

With hony they are laid upon the biting of mad dogs; being applied to the temples with vineger or oile of Roses, they take away the head-ache. They are also good against the cough and shortnesse of winde.

PEACH TREE

The Peach tree is a tree of no great bignesse: it sendeth forth divers boughes, which be so brittle, as oftentime they are broken with the weight of the fruit or with the winde. The leaves be long, nicked in the edges, like almost to those of the Walnut tree, and in taste bitter: the floures be of a light purple colour. The fruit or Peaches be round, and have as it were a chinke or cleft on the one side; they are covered with a soft and thin downe or hairy cotton, being white without, and of a pleasant taste; in the middle whereof is a rough or rugged stone, wherein is contained a kernell like unto the Almond; the meate about the stone is of a white color. The root is tough and yellowish.

The red Peach tree is likewise a tree of no great bignesse: it also sendeth forth divers boughes or branches which be very brittle. The leaves be long, and nicked in the edges like to the precedent. The floures be also like unto the former; the fruit or Peaches be round, and of a red colour on the outside; the meate likewise about the stone is of a gallant red colour. These kindes of Peaches are very like to wine in taste, and therefore marvellous pleasant.

The White Peach

Persica præcocia, or the d'avant Peach tree is like unto the former, but his leaves are greater and larger. The fruit or Peaches be of a russet colour on the one side, and on the other side next unto the Sun of a red colour, but much greater than the red Peach: the stones whereof are like unto the former: the pulpe or meate within is of a golden yellow colour, and of a pleasant taste.

Persica lutea, or the yellow Peach tree is like unto the former in leaves and flours, his fruit is of a yellow color on the out side, and likewise on the in side, harder than the rest: in the middle of the Peach is a wooddy hard and rough stone full of crests and gutters, in which doth ly a kernel much like to that of the almond, and with such a like skin: the substance within is white, and of taste somewhat bitter. The fruit hereof is of greatest pleasure, and of best taste of all the other of his kinde; although there be found at this day divers other sorts that are of very good taste, not remembred of the antient, or set down by the later Writers, whereof to speake particularly would not bee great to our pretended purpose, considering wee hasten to an end.

They are set and planted in gardens and Vineyards. I have them all in my garden, with many other sorts. The Peach tree soone comes up, it beares fruit the third or fourth yeare after it is planted, and it soon decayeth, being not of long continuance.

Maple tree

The great Maple is a beautifull and high tree, with a barke of a meane smoothnesse: the substance of the wood is tender and easie to worke on; it sendeth forth on every side very many goodly boughes and branches, which make an excellent shadow against the heat of the Sun; upon which are great, broad, and cornered leaves, much like to those of the Vine, hanging by long reddish stalkes; the floures hang by clusters, of a whitish greene colour; after them commeth up long fruit fastened together by couples, one right against another, with kernels bumping out neere to the place in which they are combined: in all the other parts flat and thin like unto parchment, or resembling the innermost wings of grashoppers: the kernels be white and little.

The great Maple is a stranger in England, onely it groweth in the walkes and places of pleasure of noble men, where it especially is planted for the shadow sake, and under the name of Sycomore tree.

THE OKE

The common Oke groweth to a great tree; the trunke or body wherof is covered over with a thicke rough barke full of chops or rifts: the armes or boughes are likewise great, dispersing themselves farre abroad: the leaves are bluntly indented about the edges, smooth, and of a shining greene colour, whereon is often found a most sweet dew and somewhat clammie, and also a fungous excrescence, which wee call Oke Apples. The fruit is long, covered with a browne, hard, and tough pilling, set in a rough scaly cup or husk: there is often found upon the body of the tree, and also upon the branches, a certaine kind of long white mosse hanging downe from the same: and sometimes another wooddie plant, which we call Missel-toe, being either an excrescence or outgrowing from the tree it selfe, or of the doung (as it is reported) of a bird that hath eaten a certaine berry.

The Oke doth scarcely refuse any ground; for it groweth in a dry and barren soile, yet doth it prosper better in a fruitfull ground; it groweth upon hills and mountaines, and likewise in vallies: it commeth up every where in all parts of England, but it is not so common in other of the South and hot regions.

The Oke doth cast his leaves for the most part about the end of Autumne: some

The Oke with his Acornes

keepe their leaves on, but dry all Winter long, untill they be thrust off by the new Spring.

Acornes if they be eaten are hardly concocted, they yeeld no nourishment to mans body, but that which is grosse, raw, and cold. Swine are fatted herewith, and by feeding thereon have their flesh hard and sound.

The decoction of Oke Apples steeped in strong white wine vineger, with a little pouder of Brimstone, and the root of *Ireos* mingled together, and set in the Sun by the space of a moneth, maketh the hair blacke, consumeth proud and superfluous flesh, taketh away sun-burning, freckles, spots, the morphew, with all deformities of the face, being washed therewith.

The Oke Apples being broken in sunder about the time of their withering, doe foreshew the sequell of the yeare, as the expert Kentish husbandmen have observed by the living things found in them: as if they finde an Ant, they foretell plenty of graine to ensue: if a white worme like a Gentill or Magot, then they prognosticate murren of beasts and cattell; if a spider, then (say they) we shall have a pestilence or some such like sickenesse to follow amongst men: these things the learned also have observed and noted; for *Matthiolus* writing upon *Dioscorides* saith, that before they have an hole through them, they containe in them either a flie, a spider, or a worme; if a flie then warre insueth, if a creeping worme, then scarcitie of victuals; if a running spider, then followeth great sicknesse or mortalitie.

Mushrumes, or Toadstooles

Some Mushrumes grow forth of the earth; other upon the bodies of old trees, which differ altogether in kindes. Many wantons that dwell neere the sea, and have fish at will, are very desirous for change of diet to feed upon the birds of the mountaines; and such as dwell upon the hills or champion grounds, do long after sea fish; many that have plenty of both, do hunger after the earthy excrescences, called Mushromes: whereof some are very venomous and full of poyson, others not so noisome; and neither of them very wholesome meate; wherefore for the avoiding of the venomous quality of the one, and that the other which is lesse venomous may be discerned from it, I have thought good to set forth their figure.

Ground Mushrums grow up in one night, standing upon a thicke and round stalke, like unto a broad hat or buckler, of a very white colour until it begin to wither, at what time it loseth his faire white, declining to yellownesse: the lower side is somewhat hollow, set or decked with fine gutters, drawne along from the middle centre to the circumference or round edge of the brim.

All Mushroms are without pith, rib, or veine: they differ not a little in bignesse and colour, some are great, and like a broad brimmed hat; others smaller, about the big-nesse of a silver coine called a

Common Mushrums to be eaten

doller: most of them are red underneath; some more, some lesse; others little or nothing red at all: the upper side which beareth out, is either pale or whitish, or else of an ill-favoured colour like ashes (they commonly call it Ash-colour) or else it seemeth to be somewhat yellow.

The Mushrums or Toodstooles which grow upon the trunkes or bodies of old trees, very much resembling

Auricula Judæ, that is, Jewes eare, doe in continuance of time grow unto the substance of wood, which the Fowlers doe call Touchwood, and are for the most halfe circuled or halfe round, whose upper part is somewhat plaine, and sometimes a little hollow, but the lower part is plaited or pursed together. This kinde of Mushrum is full of venome or poyson, especially those which grow upon the Ilex, Olive and Oke trees.

There is likewise a kinde of Mushrum called *Fungus Favaginosus*, growing up in moist and shadowie woods, which is also venomous, having a thicke and tuberous stalke, an handfull high, of a duskish colour; the top whereof is compact of many small divisions, like unto the hony combe.

Fusse balls, Pucke Fusse, and Bulfists, with which in some places of England they use to kill or smolder their Bees, when they would drive the Hives, and bereave the poore Bees of their meat, houses and lives: these are also used in some places where neighbours dwell far asunder, to carry and reserve fire from place to place, wherof it tooke the name, *Lucernarum Fungus*: in forme they are very round, sticking and cleaving unto the ground, without any stalks or stems; at the first white, but afterwards of a duskish colour, having no hole or breach in them, whereby a man may see into them, which being trodden upon doe breath forth a most thin and fine pouder, like unto smoke, very noisome and hurtfull unto the eies, causing a kinde of blindnesse, which is called Poor-blinde, or Sandblinde.

Mushrums come up about the roots of trees, in grassie places of medowes, and Ley Land newly turned; in woods also where the ground is sandy, but yet dankish: they grow likewise out of wood, forth of the rotten bodies of trees, but they are unprofitable and nothing worth. Poisonsome mushroms, as *Dioscorides* saith, groweth where old rusty iron lieth, or rotten clouts, or neere to serpents dens, or roots of trees that bring forth venomous fruit. Divers esteeme those for the best which grow in medowes, and upon mountaines and hilly places, as *Horace* saith,

The medow Mushroms are in kinde the best;
It is ill trusting any of the rest.

Galen affirmes, that they are all very cold and moist, and therefore to approach unto a venomous and murthering facultie, and ingender a clammy, pituitous, and cold nutriment if they be eaten. To conclude, few of them are good to be eaten, and most of them do suffocate and strangle the eater. Therefore I give my advice unto those that love such strange and new fangled meates, to beware of licking honey among thornes, least the sweetnesse of the one do not countervaile the sharpnesse and pricking of the other.

Hasell tree

The Hasell tree groweth like a shrub or small tree, parted into boughes without joints, tough and pliable: the leaves are broad, greater and fuller of wrinckles than those of the Alder tree, cut in the edges like a saw, of colour greene, and on the backside more white, the barke is thin: the root is thicke, strong, and growing deepe; in stead of floures hang downe catkins, aglets, or blowings, slender, and well compact: after which come the Nuts standing in a tough cup of a greene colour and jagged at the upper end, like almost to the beards in Roses. The shell is smooth and wooddy: the kernell within consisteth of a white, hard, and sound pulpe, and is covered with a thin skin, oftentimes red, most commonly white; this kernell is sweet and pleasant unto the taste.

Corylus sylvestris is our hedge Nut or Hasell Nut tree, which is very well knowne, and therefore needeth not any description: whereof there are also sundry sorts, some

great, some little, some rathe ripe, some later, as also one that is manured in our Gardens, which is very great, bigger than any Filberd, and yet a kinde of Hedge Nut: this then that hath beene said shall suffice for Hedge-Nuts.

The Hasell trees doe commonly grow in Woods, and dankish untoiled places: they are also set in Orchards, the Nuts whereof are better, and of a sweeter taste, and be most commonly redde within.

The catkins or Aglets come forth very timely, before winter be fully past, and fall away in March or Aprill, so soone as the leaves come forth.

The Filberd Nut

Beech tree

The Beech is an high tree, with boughes spreading oftentimes in manner of a circle, and with a thicke body having many armes: the barke is smooth: the timber is white, hard, and very profitable: the leaves be smooth, thin, broad, and lesser than those of the blacke Poplar: the catkins or blowings be also lesser and shorter than those of the Birch tree and yellow: the fruit or Mast is contained in a huske or cup that is prickly, and rough bristled, yet not so much as that of the Chestnut: which fruit being taken forth of the shells or urchin huskes, be covered with a soft and smooth skin like in colour and smoothnesse to the Chestnuts, but they be much lesser, and of another forme, that is to say, triangled or three cornered: the kernell within is sweet, with a certaine astriction or binding qualitie: the roots be few, and grow not deepe, and little lower than under the turfe.

The Beech tree loveth a plaine and open country, and groweth very plentifully in many Forrests and desart places of Sussex, Kent, and sundry other countries.

The Beech floureth in Aprill and May, and the fruit is ripe in September, at what time the Deere do eate the same very greedily, as greatly delighting therein; which hath caused forresters and huntsmen to call it Buck-mast.

The leaves of Beech are very profitably applied unto hot swellings, blisters, and excoriations; and being chewed they are good for chapped lips, and paine of the gums.

The kernels or mast within are reported to ease the paine of the kidnies if they be eaten. With these, mice and Sqirrels are greatly delighted, who do mightily increase by feeding thereon: Swine also be fatned herewith, and certaine other beasts: also Deere doe feed thereon very greedily: they be likewise pleasant to Thrushes and Pigeons.

Petrus Crescentius writeth, That the ashes of the wood is good to make glasse with.

The water that is found in the hollownesse of Beeches cureth the naughty scurfe, tetters, and scabs of men, horses, kine, and sheepe, if they be washed therewith.

WALL-NUT TREE

This is a great tree with a thicke and tall body: the barke is somewhat greene, and tending to the colour of ashes, and oftentimes full of clefts: the boughes spread themselves far abroad: the leaves consist of five or six fastned to one rib, like those of the ash tree, and with one standing on the top, which bee broader and longer than the particular leaves of the Ash, smooth also, and of a strong smell: the catkins or aglets come forth before the Nuts: these Nuts doe grow hard to the stalke of the leaves, by couples, or by three & three; which at the first when they be yet but tender have a sweet smel, and be covered with a green huske: under that is a wooddy shell in which the kernell is contained, being covered with a thin skin, parted almost into foure parts with a wooddy skin as it were: the inner pulpe whereof is white, sweet and pleasant to the tast; and that is when it is new gathered, for after it is dry it becommeth oily and rancke.

The Walnut tree groweth in fields neere common highwayes, in a fat and fruitfull ground, and in orchards: it prospereth on high fruitfull bankes, it loveth not to grow in watery places.

The greene and tender Nuts boyled in Sugar and eaten as Suckad, are a most pleasant and delectable meat, comfort the stomacke, and expell poison.

The oyle of Walnuts made in such manner as oyle of Almonds, maketh smooth the hands and face.

Milke made of the kernels, as Almond milke is made, cooleth and pleaseth the appetite of the languishing sicke body.

With onions, salt and honey, they are good against the biting of a mad dog or man, if they be laid upon the wound.

The Walnut Tree

251

Chestnut Tree

The Chestnut tree is a very great and high tree: it casteth forth very many boughes: the body is thicke, and sometimes of so great a compasse as that two men can hardly fathom it: the timber or substance of the wood is sound and durable: the leaves bee great, rough, wrinkled, nicked in the edges, and greater than the particular leaves of the Walnut tree. The blowings or catkins be slender, long, and greene: the fruit is inclosed in a round rough and prickly huske like to an hedge-hog or Urchin, which opening it selfe doth let fall the ripe fruit or Nut. This nut is not round, but flat on the one side, smooth, and sharpe pointed: it is covered with a hard shell, which is tough and very smooth, of a darke browne colour: the meate or inner substance of the nut is hard and white, and covered with a thin skin which is under the shell.

The Horse Chestnut groweth likewise to be a very great tree, spreading his great and large armes or branches far abroad, by which meanes it maketh a very good coole shadow. These branches are garnished with many beautifull leaves, cut or divided into five, six, or seven sections or divisions like to the Cinkefoile, or rather like the leaves of *Ricinus*, but bigger. The floures grow at the top of the stalkes, consisting of foure small leaves like the Cherry blossome, which turne into round rough prickly heads like the former, but more sharpe and harder: the Nuts are also rounder.

The first growes on mountaines and shadowie places, and many times in the vallies: they love a soft and blacke soile. There be sundry woods of Chestnuts in England, as a mile and a halfe from Feversham in Kent, and in sundry other places: in some countries they be greater and pleasanter: in others smaller, and of worse taste. The Horse Chestnut groweth in Italy, and in sundry places of the East-countries.

The blowings or aglets come forth with the leaves in Aprill; but the Nuts later, and be not ripe till Autumne.

The Horse Chestnut is called in Latine, *Equina Castanea*: in English, Horse Chestnut, for that the people of the East countries do with the fruit thereof cure their horses of the cough, shortnesse of breath, and such like diseases.

Our common Chestnuts are very dry and binding, and be neither hot nor cold, but in a mean betweene both: yet have they in them a certaine windinesse, and by reason of this, unlesse the shell be first cut, they skip suddenly with a cracke out of the fire whilest they be rosting.

ELDER TREE

There be divers sorts of Elders, some of the land, and some of the water or marish grounds.

The common Elder groweth up now and then to the bignesse of a mean tree, casting his boughes all about, and oftentimes remaineth a shrub: the body is almost all wooddy, having very little pith within; but the boughs, and especially the yong ones, which be jointed, are full of pith within, and have but little wood without: the barke of the body and great armes is rough and full of chinks, and of an ilfavoured wan colour like ashes: that of the boughs is not very smooth, but in colour almost like; and that is the outward barke; for there is another under it neerer to the wood, of colour green: the substance of the wood is sound, somwhat yellow, and that may be easily cleft: the leaves consist of five or six particular ones fastned to one ribbe, like those of the Walnut tree, but every particular one is lesser, nicked in the edges, and of a ranke and stinking smell. The floures grow on spoky rundles, which be thin and scattered, of a white colour and sweet smell: after them grow up little berries, green at

253

the first, afterwards blacke, whereout is pressed a purple juice, which being boiled with Allom and such like things, doth serve very well for the Painters use, as also to colour vineger: the seeds in these are a little flat and somwhat long.

Elder Tree

There groweth oftentimes upon the bodies of those old trees or shrubs a certaine excrescence called *Auricula Judæ* or Jewes eare, which is soft, blackish, covered with a skin, somewhat like now and then to a mans eare, which being plucked off and dried, shrinketh together and becometh hard.

This Elder groweth every where: it is planted about Cony-boroughs for the shadow of the Conies.

These kindes of Elders floure in Aprill and May, and their fruit is ripe in September.

The seeds contained within the berries dried are good for such as have the Dropsie, and such as are too fat and would faine be leaner, if they be taken in a morning to the quantitie of a dram with wine for a certain space.

The green leaves pouned with Deeres suet or Bulls tallow, are good to be laid to hot swellings and tumors, and doe asswage the pain of the gout.

The gelly of the Elder, otherwise called Jewes eare, hath a binding and drying qualitie: the infusion thereof, in which it hath bin steeped a few houres, taketh away inflammations of the mouth and throat, if they be washed therewith, and doth in like manner help the uvula.

The Elder Rose groweth like an hedge tree: the leaves are like the vine leaves, among which come forth goodly floures of a white colour, sprinkled and dashed heere and there with a light and thin carnation colour, and do grow thicke and closely compact together, of great beauty. In my garden there groweth not any fruit upon this tree, nor in any other place, for ought that I can understand. It is called in English, Gelders Rose.

WHITE SATTIN FLOURE

White Sattin

Bolbonac or the Sattin floure hath hard and round stalks, dividing themselves into many other small branches, beset with leaves like Dames Violets or Queenes Gillofloures, somewhat broad, and snipt about the edges, and in fashion almost like Sauce alone, or Jacke by the hedge, but that they are longer and sharper pointed. The stalks are charged or loden with many floures like the common stocke Gillofloure, of a purple colour: which being fallen, the seed comes forth, contained in a flat thin cod, with a sharp point or pricke at one end, in fashion of the Moon, but somewhat blackish. This cod is composed of three filmes or skins, whereof the two outmost are of an overworne ash colour, and the innermost or that in the middle, wheron the seed doth hang or cleave, is thinne and cleere shining, like a shred of white

255

Sattin newly cut from the piece. The whole plant dieth the same yeare that it hath borne seed, and must be sown yearly. The root is compact of many tuberous parts like Key clogs, or like the great Asphodill.

These Plants are set and sowne in gardens, notwith-standing one hath bin found wild in the woods about Pinner and Harrow on the hill 12 miles from London, and in Essex likewise about Horn-church.

We call this herbe in English, Penny floure, or Mony-floure, Silver Plate, Pricksong-wort; in Norfolke, Sattin, and white Sattin; and among our women it is called Honestie.

The seed of Bulbonac is sharpe of taste, like in force to the seed of Treacle mustard: the roots likewise are somewhat of a biting qualitie, but not much: they are eaten with sallads as certaine others root are.

A certain Helvetian Surgeon composed a most singular unguent for green wounds, of the leaves of Bolbonac and Sanicle stamped together, adding thereto oile and wax. The seed is greatly commended against the falling sicknesse.

TRUE SAFFRON

The floure of Saffron doth first rise out of the ground nakedly in September, and his long smal grassie leaves shortly after, never bearing floure and leafe at once. The floure consisteth of six small blew leaves tending to purple, having in the middle many small yellow strings or threds; among which are two, three, or more thicke fat chives of a fierie colour somewhat reddish, of a strong smell when they be dried, which doth stuffe and trouble the head.

Common or best knowne Saffron groweth plentifully in Cambridge-shire, Saffron-Walden, and other places thereabout, as corne in the fields. Saffron beginneth to floure in September, and presently after spring up the leaves, and remaine greene all the Winter long.

Saffron is called in Latine, *Crocus*: in Mauritania, *Saffaran*: in Spanish, *Açafron*: in English, Saffron: in the Arabicke tongue, *Zahafaran*.

Avicen affirmeth, That it causeth head-ache, and is hurtfull to the braine, which it cannot do by taking it now and then, but by too much using of it; for the too much using of it cutteth off sleep, through want whereof the head and sences are out of frame. But the moderat use thereof is good for the head, and maketh the sences more quicke and lively, shaketh off heavy and drowsie sleepe, and maketh a man merry.

Also Saffron strengthneth the heart, concocteth crude and raw humors of the chest, opens the lungs, and removeth obstructions.

It is also such a speciall remedie for those that have consumption of the lungs, and are, as wee terme it, at deaths doore, and almost past breathing, that it bringeth breath again, and prolongeth life for certaine dayes, if ten or twenty graines at the most be given with new or sweet Wine. For we have found by often experience, that being taken in that sort, it presently and in a moment removeth away difficulty of breathing, which most dangerously and suddenly hapneth.

The eyes being anointed with the same dissolved in milke or fennel or rose water, are preserved from being hurt by the small pox or measels, and are defended thereby from humors that would fal into them.

The chives steeped in water serve to illumine or (as we say) limne pictures and imagerie, as also to colour sundry meats and confections.

The weight of ten grains of Saffron, the kernels of Walnuts two ounces, Figs two ounces, Mithridate one dram, and a few Sage leaves stamped together with a sufficient quantitie of Pimpernel water, and made into a masse or lumpe, and kept in a glasse for your use, and thereof 12 graines given in the morning fasting, preserveth from the pestilence, and expelleth it from those that are infected.

MEDOW SAFFRON

Mede Saffron

There be sundry sorts of medow Saffrons, differing very notably as well in the colour of their floures, as also in nature and country from whence they had their being, as shall be declared.

Medow Saffron hath three or four leaves rising immediately forth of the ground, long, broad, smooth, fat, much like to the leaves of the white Lillie in forme and smoothnesse: in the middle whereof spring up three or foure thicke cods of the bignesse of a small Wall-nut, standing upon short tender foot-stalkes, three square, and opening themselves when they be ripe, full of seed something round, and of a blackish red colour: and when this seed is ripe, the leaves together with the stalkes doe fade and fall away. In September the floures bud forth, before any leaves appeare, standing upon short tender and whitish stemmes, like in forme and colour to the floures of Saffron, having in the middle small chives or threds of a pale yellow colour, altogether unfit for meat or medicine.

The second kinde of Mede Saffron is like the precedent, differing onely in the colour of the floures, for that this plant doth bring forth white leaves, which of some hath beene taken for the true *Hermodactylus*; but in so doing they have committed the greater error.

The third kinde bringeth forth his leaves in the Spring of the yeare, thicke, fat, shining, and smooth, not unlike the leaves of Lillies, which doe continue greene unto the end of June; at which time the leaves do wither away, but in the beginning of September there shooteth forth of the ground naked milke white floures without any greene leafe at all: but so soone as the Plant hath done bearing of floures, the root remaines in the ground, not sending forth any thing untill Februarie in the yeare following.

Divers name it in Latine *Bulbus agrestis*, or wild Bulbe: in French, *Mort au chien*. Some have taken it to be the true Hermodactyl, yet falsely. Other some call it *Filius ante patrem*, although there is a kinde of *Lysimachia* or Loose-strife so called, because it first bringeth forth his long cods with seed, and then the floure after, or at the same time at the end of the said cod. But in this Mede Saffron it is far otherwise, because it bringeth forth leaves in Februarie, seed in May, and floures in September; which is a thing cleane contrarie to all other plants whatsoever, for that they doe first floure, and after seed: but this Saffron seedeth first, and foure moneths after brings forth flowers: and therefore some have thought this a fit name for it, *Filius ante Patrem*: and we accordingly may call it, The Sonne before the Father.

BRAMBLE OR BLACKE-BERRY BUSH

The common Bramble bringeth forth slender branches, long, tough, easily bowed, ramping among hedges and whatsoever stands neere unto it; armed with hard and sharpe prickles, whereon doe grow leaves consisting of many set upon a rough middle ribbe, greene on the upper side, and underneath somewhat white: on the tops of the stalkes stand certaine floures, in shape like those of the Brier Rose, but lesser, of colour white, and sometimes washt over with a little purple: the fruit or berry is like that of the Mulberry, first red, blacke when it is ripe, in taste betweene sweet and soure, very soft, and full of grains: the root creepeth, and sendeth forth here and there young springs.

The Bramble Bush

The Raspis or Framboise bush hath leaves and branches not much unlike the common Bramble, but not so rough nor prickly, and sometimes without any prickles at all, having onely a rough hairinesse about the stalkes: the fruit in shape and proportion is like those of the Bramble, red when they be ripe, and covered over with a little downinesse; in taste not very pleasant. The root creepeth far abroad, whereby it greatly encreaseth.

The Bramble groweth for the most part in every hedge and bush.

The Raspis is planted in gardens: it groweth not wilde that I know of, except in the field by a village in Lancashire called Harwood, not far from Blackeburne. I found it among the bushes of a causey, neere unto a village called Wisterson, where I went to schoole, two miles from the Nantwich in Cheshire.

These floure in May and June with the Roses: their fruit is ripe in the end of August and September.

The Bramble is called in Latine, *Rubus*, and *Sentis*, and *Vepres*, as *Ovid* writeth in the first booke of Metamorphosis:

Or to the Hare, that under Bramble closely lying, spies
The hostile mouth of Dogs.——

It is called in French, *Rouce*: in English, Bramble bush, and Blacke-berry bush.

The Raspis is called in Latine, *Rubus Idæa*, of the mountaine Ida on which it groweth: in English, Raspis, Framboise, and Hinde-berry.

The yong buds or tender tops of the Bramble bush, the floures, the leaves, and the unripe fruit, being chewed, stay all manner of bleedings. They heale the eies that hang out.

The ripe fruit is sweet, and containeth in it much juyce of a temperate heate, therefore it is not unpleasant to be eaten.

The leaves of the Bramble boyled in water, with honey, allum, and a little white wine added thereto, make a most excellent lotion or washing water, and the same decoction fastneth the teeth.

MULBERRIE TREE

The common Mulberrie tree is high, and ful of boughes: the body wherof is many times great, the barke rugged; and that of the root yellow: the leaves are broad and sharpe pointed, something hard, and nicked on the edges; in stead of floures, are blowings or catkins, which are downy: the fruit is long, made up of a number of little graines, like unto a black-Berrie, but thicker, longer, and much greater, at the first greene, and when it is ripe blacke, yet is the juyce (whereof it is full) red: the root is parted many waies.

The white Mulberrie tree groweth untill it be come unto a great and goodly stature, almost as big as the former: the leaves are rounder, not so sharpe pointed, nor so deeply snipt about the edges, yet sometimes sinuated or deeply cut in on the sides, the fruit is like the former, but that it is white and somewhat more tasting like wine.

The Mulberrie trees grow plentifully in Italy and other hot regions, where they doe maintaine great woods and groves of them, that their Silke wormes may feed thereon. The Mulberry tree is fitly set by the slip; it may also be grafted or inoculated into many trees, being grafted in a white Poplar, it bringeth forth white Mulberries, as *Beritius* in his Geoponickes reporteth. These grow in sundry gardens in England.

Of all the trees in the Orchard the Mulberry doth last bloome, and not before the cold weather is gone in May (therefore the old Writers were wont to call it the wisest tree) at which time the Silke wormes do seeme to revive, as having then wherewith to feed and nourish themselves, which all the winter before do lie like small graines or seeds, as knowing their proper times both to performe their duties for which they were created, and also when they may have wherewith to maintaine and preserve their owne bodies, unto their businesse aforesaid.

Mulberries are good to quench thirst, they stir up an appetite to meate, they are not hurtfull to the stomacke, but they nourish the body very little, being taken in the second place, or after meate.

The barke being steeped in vineger helpeth the tooth-ache: of the same effect is also the decoction of the leaves and barke, saith *Dioscorides*, who sheweth that about harvest time there issueth out of the root a juyce, which the next day after is found to be hard, and that the same is very good against the tooth-ache.

THE OLIVE TREE

The tame or manured Olive tree groweth high and great with many branches, full of long narrow leaves not much unlike the leaves of Willowes, but narrower and smaller: the floures be white and very small, growing upon clusters or bunches: the fruit is long and round, wherein is an hard stone; from which fruit is pressed that liquor which we call oyle Olive.

The wilde Olive is like unto the tame or garden Olive tree, saving that the leaves are something smaller; among which sometimes doe grow many prickly thornes: the fruit hereof is lesser than of the former, and moe in number, which do seldome come to maturity or ripenes in so much that the oile which is made of those berries, continueth ever greene, and is called oile Omphacine, or oile of unripe Olives.

Both the tame and the wilde Olive trees grow in very many places of Italy, France, and Spaine, and also in the Islands adjoyning: they are reported to love the sea coasts; for most doe thinke, as *Columella* writeth, that above sixty miles from the sea they either die, or else

The manured Olive tree

bring forth no fruit: but the best, and they that doe yeeld the most pleasant oile are those that grow in the Island called Candy.

All the Olive trees floure in the moneth of June: the fruit is gathered in November or December: when they be a little dried and begin to wrinkle they are put into the presse, and out of them is squeezed oyle, with water added in the pressing: the Olives which are to bee preserved in salt and pickle must be gathered before they be ripe, and whilest they are greene.

The juyce is pressed forth of the stamped leaves, with Wine added thereto (which is better) or with water, and being dried in the Sun it is made up into little cakes like perfumes.

The oile of ripe Olives mollifieth and asswageth paine, dissolveth tumors or swellings, is good for the stiffenesse of the joynts, and against cramps, especially being mingled according to art, with good and wholesome herbes appropriate unto those diseases and griefes, as *Hypericon*, Cammomill, Dill, Lillies, Roses, and many others, which do fortifie and increase his vertues.

QUINCE TREE

The Quince tree is not great, but growes low, and many times in maner of a shrub: it is covered with a rugged barke, which hath on it now and then certain scales: it spreadeth his boughes in compasse like other trees, about which stand leaves somewhat round like those of the common Apple tree, greene and smooth above, and underneath soft and white: the flours be of a white purple colour: the fruit is like an Apple, saving that many times it hath certain embowed & swelling divisions: it differeth in fashion and bignesse; for some Quinces are lesser and round, trust up together at the top with wrinckles, others longer and greater: the third sort be of a middle manner betwixt both; they are all of them set with a thinne cotton or freese, and be of the colour of gold, and hurtfull to the head by reason of their strong smell; they all likewise have a kinde of choking tast: the pulp within is yellow, and the seed blackish, lying in hard skins as do the kernels of other apples.

The Quince groweth in gardens and orchards, and is planted oftentimes in hedges and Fences belonging to

Gardens and Vineyards: it delighteth to grow on plain and even grounds, and somwhat moist withall.

The Marmalad or Cotiniat made of quinces and sugar is good and profitable to strengthen the stomack, that it may retain and keep the meat therein untill it be perfectly digested. Which Cotiniat is made in this manner:

Take faire Quinces, paire them, cut them in pieces, and cast away the core, then put unto every pound of Quinces a pound of Sugar, and to every pound of Sugar a pinte of water: these must be boiled together over a stil fire till they be very soft, then let it be strained or rather rubbed through a strainer or an hairy Sive, which is better, and then set it over the fire to boile againe, untill it be stiffe, and so box it up, and as it cooleth put thereto a little Rose water, and a few graines of muske mingled together, which will give a goodly taste to the Cotiniat.

This is the way to make Marmalad. Take whole Quinces and boile them in water until they be as soft as a scalded codling or apple, then pill off the skin, and cut off the flesh, and stamp it in a stone morter, then straine it as you did the Cotiniat; afterward put it in a pan to dry, but not to seeth at all, and unto every pound of the flesh of quinces put three quarters of a pound of sugar, and in the cooling you may put in rose water and a little muske, as was said before.

Many other excellent dainty and wholesome Confections are to be made of Quinces, as jelly of Quinces, and such like conceits, which for brevities sake I do now let passe.

TURKIE CORNE

The stalk of Turky wheat is like that of the Reed. The fruit is contained in verie big eares, covered with coats and filmes like husks & leaves, as if it were a certain sheath. The seeds are great, of the bignesse of common peason.

These kinds of grain were first brought into Spaine, and then into other provinces of Europe: not (as some suppose) out of Asia *minor*, which is the Turks dominions; but out of America and the Islands adjoining, as out of Florida, and Virginia or Norembega, where they use to sow or set it to make bread of it, where it growes much higher than in other countries. It is planted in the gardens of these Northern regions, where it commeth to ripenesse when the summer falleth out to be faire and hot; as my selfe have seen by proof in myne owne garden.

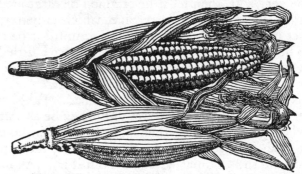

Turky Wheat in the huske, as also naked or bare

It is sowen in these countries in March and Aprill, and the fruit is ripe in September.

Turky wheat doth nourish far lesse than either wheat, rie, barly, or otes. The bread which is made thereof is meanely white, without bran: it is hard and dry as Bisket is, and hath in it no clamminesse at all; for which cause it is of hard digestion, and yeeldeth to the body little or no nourishment. Wee have as yet no certaine proofe or experience concerning the vertues of this kinde of Corne; although the barbarous Indians, which know no better, are constrained to make a vertue of necessitie, and thinke it a good food: whereas we may easily judge, that it nourisheth but little, and is of hard and evill digestion, a more convenient food for swine than for man.

POTATO'S

This plant (which is called of some Skyrrets of Peru) is generally of us called Potatus or Potato's. It hath long rough flexible branches trailing upon the ground like unto those of Pompions, whereupon are set greene three cornered leaves very like those of the wilde Cucumber.

The roots are many, thicke, and knobby, like unto the roots of Peonies, or rather of the white Asphodill, joined together at the top into one head, in maner of the Skyrret, which being divided into divers parts and planted, do make a great increase, especially if the greatest roots be cut into divers goblets, and planted in good and fertile ground.

The Potato's grow in India, Barbarie, Spaine, and other hot regions; of which I planted divers roots (which I bought at the Exchange in London) in my garden, where they flourished until winter, at which time they perished and rotted.

The Potato roots are among the Spaniards, Italians, Indians, and many other nations, ordinarie and common meat; which no doubt are of mighty and nourishing parts, and doe strengthen and comfort nature; whose nutriment is as it were a mean between flesh and fruit, but somewhat windie; yet being rosted in the embers they lose much of their windinesse, especially being eaten sopped in wine.

Of these roots may be made conserves no lesse toothsome, wholesome, and dainty, than of the flesh of Quinces; and likewise those comfortable and delicate meats called in shops, *Morselli, Placentulæ*, and divers other such like.

These roots may serve as a ground or foundation whereon the cunning Confectioner or Sugar-Baker may worke and frame many comfortable delicat Conserves and restorative sweet-meats.

They are used to be eaten rosted in the ashes. Some when they be so rosted infuse and sop them in wine: and

267

others to give them the greater grace in eating, do boile them with prunes and so eat them: likewise others dresse them (being first rosted) with oile, vineger, and salt, every man according to his owne taste and liking. Notwithstanding howsoever they be dressed, they comfort, nourish, and strengthen the body.

POTATO'S OF VIRGINIA

Virginian Potato hath many hollow flexible branches trailing upon the ground, three square, uneven, knotted

Virginian Potatoes

or kneed in sundry places at certaine distances: from the which knots commeth forth one great leafe made of divers leaves, some smaller, and others greater, set together upon a fat middle rib by couples, of a swart

268

greene colour tending to rednesse; the whole leafe resembling those of the Winter-Cresses, but much larger; in taste at the first like grasse, but afterward sharp and nipping the tongue. From the bosome of which leaves come forth long round slender footstalkes, whereon grow very fair and pleasant floures.

It groweth naturally in America, where it was first discovered, as reporteth *Clusius*, since which time I have received roots hereof from Virginia, otherwise called Norembega, which grow & prosper in my garden as in their owne native country.

The Indians call this plant *Pappus*, meaning the roots; by which name also the common Potatoes are called in those Indian countries. Wee have it's proper name mentioned in the title.

The vertues be referred to the common Potato's, being likewise a food, as also a meat for pleasure, equall in goodnesse and wholesomnesse to the same, being either rosted in the embers, or boiled and eaten with oile, vineger and pepper, or dressed some other way by the hand of a skilfull Cooke.

THE MANURED VINE

The trunke or body of the Vine is great and thicke, very hard, covered with many barks, which are full of cliffes or chinks; from which grow forth branches as it were armes, many wayes spreading; out of which come forth jointed shoots or springs; and from the bosom of those joints, leaves and clasping tendrels, and likewise bunches or clusters full of grapes: the leaves be broad, something round, five cornered, and somewhat indented about the edges: amongst which come forth many clasping tendrels, that take hold of such props or staies as stand next unto it. The grapes differ both in colour and greatnesse, and also in many other things, which to distinguish severally were impossible, considering the infinite sorts or kinds, and also those which are transplanted from one region or clymat to another, do likewise alter both from the forme and taste they had before: wherefore it shall be sufficient to set forth the figure of the manured grape, and speak somwhat of the rest.

There be some Vines that bring forth grapes of a whitish or reddish yellow colour; others of a deep red, both in the outward skin, juice, and pulpe within.

There be others whose grapes are of a blew colour, or something red, yet is the juice like those of the former. These grapes doe yeeld forth a white wine before they are put into the presse, and a reddish or paller wine when they are trodden with the husks, & so left to macerate or ferment, with which if they remain too long, they yeeld forth a wine of a higher colour.

There be others which make a black and obscure red wine, whereof some bring bigger clusters, and consist of greater grapes, others of lesser; some grow more clustered or closer together, others looser; some have but one stone, others more; some make a more austere or harsh wine, others a more sweet: of some the old wine is best, of divers the first yeres wine is most excellent: some bring forth fruit foure square, of which kindes we have great plenty.

The plant that beareth those small Raisins which are commonly called Corans or Currans, or rather Raisins of Corinth, is not that plant which among the vulgar people is taken for Currans, it being a shrub or bush that brings forth small clusters or berries, differing as much as may be from Corans, having no affinitie with the Vine or any kinde thereof. The Vine that beareth small Raisins hath a body or stock as other Vines have, branches and tendrels likewise. The leaves are larger than any of the others, snipt about the edges like the teeth of a saw: among which come forth clusters of grapes in forme like the other, but smaller, of a blewish colour; which being ripe are gathered and laid upon hurdles, carpets, mats, and such like, in the Sun to dry: then they are caried to some house and laid upon heaps, as wee lay apples and corne in a garner, untill the merchants buy them: then do they put them into large buts or other woodden vessels, and

The manured Vine

tread them downe with their bare feet, which they call Stiving, and so they are brought into these parts for our use.

A wise husbandman will commit to a fat and fruitfull soile a leane Vine, of his own nature not too fruitfull: to a lean ground a fruitfull Vine: to a close and compact earth a spreading vine, and that is full of matter to make branches of: to a loose and fruitful soile a Vine of few branches.

Grapes have the preheminence among the Autumne fruits, and nourish more than they all, but yet not so much as figges; and they have in them little ill juice, especially when they bee thorow ripe.

Grapes may be kept the whole yeare, being ordered after the same manner that *Joachimus Camerarius* reporteth. You shall take, saith hee, the meale of mustard seed, and strew in the bottome of any earthen pot well leaded; whereupon you shall lay the fairest bunches of the ripest grapes, the which you shall cover with more of the foresaid meale, and lay upon it another sort of Grapes, so doing untill the pot be full: then shall you fill

271

up the pot to the brimme with a kind of sweet wine called Must. The pot being very close covered shall be set into some cellar or other cold place: the grapes you may take forth at your pleasure, washing them with faire water from the pouder.

To speake of wine the juice of Grapes, which being newly pressed forth is called *Mustum* or new wine; after the dregs and drosse are setled, and it appeareth pure and cleer, it is called in English, Wine, and that not unproperly. For certain other juices, as of Apples, Pomegranats, Peares, Medlars, Services, or such otherwise made (for examples sake) of Barley and Graine, be not at all simply called wine, but with the name of the thing added whereof they do consist. Hereupon is the wine which is pressed forth of the Pomegranate berries named *Rhoites*, or wine of Pomegranats; out of Quinces, *Cydonites*, or wine of Quinces: out of Peares, *Ápyites* or Perry; and that which is compounded of Barley is called *Zythum*, or Barley wine: in English, Ale or Beere.

And other certain wines have borrowed syrnames of the plants that have bin infused or steeped in them; and yet all wines of the Vine, as Wormwood wine, Myrtle wine, and Hyssop wine, which are all called artificiall wines.

That is properly and simply called wine which is pressed out of the grapes of the vine, and is without any manner of mixture.

The kindes of Wines are not of one nature, nor of one facultie or power, but of many, differing one from another; for there is one difference thereof in taste, another in colour, the third is referred to the consistence or substance of the Wine; the fourth consisteth in the vertue and strength thereof. *Galen* addeth that which is found in the smell, which belongs to the vertue and strength of the Wine.

It is good for such as are in a consumption, by reason

of some disease, and that have need to have their bodies nourished and refreshed (alwaies provided they have no fever,) as *Galen* saith in his seventh booke of the Method of curing. It restoreth strength most of all other things, and that speedily: It maketh a man merry and joyfull: It putteth away feare, care, troubles of minde, and sorrow: and bringeth sleepe gently.

And these things proceed of the moderate use of wine: for immoderate drinking of wine doth altogether bring the contrarie. They that are drunke are distraughted in minde, become foolish, and oppressed with a drowsie sleepinesse, and be afterward taken with the Apoplexy, the gout, or altogether with other most grievous diseases.

And seeing that every excesse is to be shunned, it is expedient most of all to shun this, by which not only the body, but also the minde receiveth hurt.

Wherefore we thinke, that wine is not fit for men that be already of full age, unlesse it be moderately taken, because it carrieth them headlong into fury and lust, and troubleth and dulleth the reasonable part of the minde.

There is drawne out of Wine a liquor, which in Latine is commonly called *Aqua vitæ*, or water of life, and also *Aqua ardens*, or burning water, which as distilled waters are drawne out of herbes and other things, is after the same manner distilled out of strong wine, that is to say, by certaine instruments made for this purpose, which are commonly called Limbeckes.

This water is good for all those that are made cold either by a long disease, or through age, as for old and impotent men: for it cherisheth and increaseth naturall heate; upholdeth strength, repaireth and augmenteth the same: it prolongeth life, quickeneth all the senses, and doth not only preserve the memory, but also recovereth it when it is lost: it sharpeneth the sight.

¶ *The briefe summe of that hath beene said of the Vine.*

Almighty God for the comfort of mankinde ordained Wine; but decreed withall, That it should be moderately taken, for so it is wholsome and comfortable: but when measure is turned into excesse, it becommeth unwholesome, and a poyson most venomous. Besides, how little credence is to be given to drunkards it is evident; for though they be mighty men, yet it maketh them monsters, and worse than brute beasts. Finally in a word to conclude; this excessive drinking of Wine dishonoreth Noblemen, beggereth the poore, and more have beene destroied by surfeiting therewith, than by the sword.

TRAVELLERS-JOY

The plant which *Lobel* setteth forth under the title of *Viorna*, *Dodonæus* makes *Vitis alba*; but not properly; whose long wooddy and viny branches extend themselves very far, and into infinite numbers, decking with his clasping tendrels and white starre-like floures (being very sweet) all the bushes, hedges, and shrubs that are neere unto it. It sends forth many branched stalkes, thicke, tough, full of shoots and clasping tendrels, wherewith it foldeth it selfe upon the hedges, and taketh hold and climeth upon every thing that standeth neere unto it. The leaves are fastned for the most part by fives upon one rib or stem, two on either side, and one in the midst or point standing alone; which leaves are broad like those of Ivy, but not cornered at all: among which come forth clusters of white floures, and after them great tufts of flat seeds, each seed having a fine white plume like a feather fastned to it, which maketh in the Winter a goodly shew, covering the hedges white all over with his feather-like tops. The root is long, tough, and thicke, with many strings fastned thereto.

The Travellers-Joy is found in the borders of fields among thornes and briers, almost in every hedge as you go from Gravesend to Canturbury in Kent; in many places of Essex, and in most of these Southerly parts about London, but not in the North of England that I can heare of.

The floures come forth in July: the beautie thereof appeares in November and December.

This plant is commonly called *Viorna, quasi vias ornans,* of decking and adorning waies and hedges, where people travel; and thereupon I have named it the Travellers-Joy.

These plants have no use in physicke as yet found out, but are esteemed onely for pleasure, by reason of the goodly shadow which they make with their thicke bushing and clyming, as also for the beauty of the floures, and the pleasant sent or savour of the same.

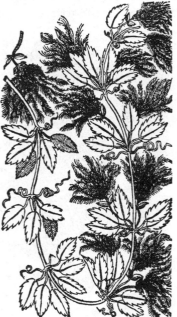

The Spanish Travellers-Joy

CEDAR TREE

The great cedar is a very big & high tree, not onely exceeding all other resinous trees and those which bear fruit like unto it, but in his talnesse and largenesse farre surmounting all other trees: the body or trunk thereof is commonly of a mighty bignesse, insomuch as four men are not able to fathom it, as *Theophrastus* writeth. The

275

bark of the lower part which proceedeth out of the earth, to the first young branches or shoots, is rough and harsh; the rest which is among the boughes is smooth and glib; the boughes grow forth almost from the bottom, and not farre from the ground, even to the very top, waxing by degrees lesser and shorter stil as they grow higher, the tree bearing the forme and shape of a Pyramide or sharpe pointed steeple: these compasse the body round about in maner of a circle, and are so orderly placed by degrees, that a man may clymbe up by them to the very top as by a ladder: the leaves be small and round like those of the Pine tree, but shorter, and not so sharp pointed: all the cones or clogs are far shorter and thicker than those of the Fir tree, compact of soft, not hard scales, which hang not downewards, but stand upright upon the boughes, whereunto also they are so strongly fastned, as they can hardly be pluckt off without breaking some part of the

The great Cedar of Libanus

branches, as *Bellonius* writeth. The timber is extreame hard, and rotteth not, nor waxeth old; there is no wormes nor rottennesse can hurt or take the hard matter or heart of this wood, which is very odoriferous and somwhat red. *Solomon* King of the Jewes did therefore build Gods Temple in Jerusalem of Cedar wood. The Gentiles were wont to make their Divels or Images of this kinde of wood, that they might last the longer.

The Cedar trees grow upon the snowy mountaines, as in Syria upon Mount Libanus, on which there remaine some even to this day, saith *Bellonius*, planted as is thought by *Solomon* himselfe: they are likewise found on the mountains Taurus and Amanus, in cold and stony places. The merchants of the Factorie of Tripolis told me, That the Cedar tree groweth upon the declining of the mount Libanus, neere to the hermitage by the city Tripolis in Syria: The inhabitants of Syria use it to make boats of, for want of the Pine tree.

The Cedar tree remaines alwaies green, as other trees which beare such maner of fruit: the timber of the Cedar tree, and the images and other works made thereof, seem to sweat and send forth moisture in moist and rainy weather, as do likewise all that have an oily juice, as *Theophrastus* witnesseth.

There issueth out of this tree a rosin like to that which issueth out of the Fir tree, very sweet in smell, of a clammy or cleaving substance, the which if you chew in your teeth it will hardly be gotten forth again, it cleaveth so fast: at the first it is liquid and white, but being dried in the Sunne it waxeth hard: if it be boiled in the fire an excellent pitch is made thereof called Cedar pitch.

The Egyptians were wont to coffin and embalme their Dead in Cedar and with Cedar pitch, although they used other means, as *Herodotus* recordeth.

CYPRESSE TREE

The tame or manured Cypresse tree hath a long thicke and straight body ; whereupon many slender branches do grow, which do not spred abroad like the branches of other trees, but grow up alongst the body, yet not touching the top: they grow after the fashion of a steeple, broad below, and narrow toward the top: the substance of the wood is hard, sound, well compact, sweet of smell, and somewhat yellow, almost like the yellow Saunders, but not altogether so yellow, neither doth it rot nor wax old, nor cleaveth or choppeth it selfe. The leaves are long, round like those of Tamariske, but fuller of substance. The fruit or nuts do hang upon the boughes, being in manner like to those of the Larch tree, but yet thicker and more closely compact: which being ripe do of themselves part in sunder, and then falleth the seed, which is shaken out with the winde: the same is small, flat, very thin, of a swart ill favoured colour, which is pleasant to Ants or Pismires, and serveth them for food.

The tame and manured Cypresse groweth in hot countries, as in Candy, Lycia, Rhodes, and also in the territory of Cyrene: it is reported to be likewise found on the hils belonging to mount Ida, and on the hills called *Leuci*, that is to say, white, the tops whereof be alwaies covered with snow. It groweth likewise in divers places of England where it hath been planted, as at Sion a place neere London, sometimes a house of Nunnes: it groweth also at Greenewich, and at other places, and likewise at Hampsted in the garden of Mr. *Wade*, one of the Clerkes of her Majesties privie Councell.

Theophrastus attributeth great honor to this tree, shewing that the roots of old Temples became famous by reason of that wood, and that the timber thereof, of which the rafters are made is everlasting, and it is not hurt there by rotting, cobweb, nor any other infirmitie or corruption.

It is reported that the smoke of the leaves doth drive away gnats, and that the clogs doe so likewise. The shavings of the wood laid among garments preserve them from the moths: the rosin killeth Moths, little wormes, and magots.

YEW TREE

The Yew Tree, as *Galen* reporteth, is of a venomous

quality, and against mans nature. *Dioscorides* writeth, and generally all that heretofore have dealt in the facultie of Herbes, that the Yew tree is very venomous to be taken inwardly, and that if any doe sleepe under the shadow thereof it causeth sicknesse and oftentimes death. Moreover, they say that the fruit thereof being eaten is not onely dangerous and deadly unto man, but if birds doe eat thereof it causeth them to cast their feathers, and many times to die. All which I dare boldly affirme is altogether untrue: for when I was yong and went to schoole, divers of my schoole-fellowes and likewise my selfe did eat our fils of the berries of this tree, and have not onely slept under the shadow thereof, but among the branches also, without any hurt at all, and that not one time, but many times. *Theophrastus* saith, That labouring beasts die, if they doe eate of the leaves; but such cattell as chew their cud receive no hurt at all thereby.

Holme, Holly, or Hulver tree

The Holly is a shrubbie plant, notwithstanding it oftentimes growes to a tree of a reasonable bignesse: the boughes whereof are tough and flexible, covered with a smooth and green barke. The substance of the wood is hard and sound, and blackish or yellowish within, which doth also sinke in the water, as doth the Indian wood which is called *Guaiacum*: the leaves are of a beautifull green colour, smooth and glib, like almost the bay leaves, but lesser, and cornered in the edges with sharp prickles, which notwithstanding they want or have few when the tree is old: the floures be white, and sweet of smell: the berries are round, of the bignesse of a little Pease, or not much greater, of colour red, of tast unpleasant, with a white stone in the midst, which do not easily fall away, but hang on the boughes a long time: the root is wooddy.

There is made of the smooth barke of this tree or shrub, Birdlime, which the birders and country men do use to take birds with: they pul off the barke, and make a ditch in the ground, specially in moist, boggy, or soggy earth, wherinto they put this bark, covering the ditch with boughes of trees, letting it remaine there til it be rooted and putrified, which will be done in the space of twelve daies or thereabout: which done, they take it forth, and beat in morters untill it be come to the thicknesse and clamminesse of Lime: lastly, that they may clear it from pieces of barke and other filthinesse, they do wash it very often: after which they adde unto it a little oyle of nuts, and after that do put it up in earthen vessels.

Holly tree

The Holly tree groweth plentifully in all countries. It groweth green both winter and summer; the berries are ripe in September, and they do hang upon the tree a long time after.

This tree or shrub is called in Latine *Agrifolium*: in French, *Hous* and *Housson*: in English, Holly, Hulver, and Holme.

The Birdlime which is made of the barke hereof is no less hurtfull than that of Misseltoe, for it is marvellous clammy, it glueth up all the intrails, and by this meanes it bringeth destruction to man, not by any quality, but by his gluing substance.

280

Ivy

Ivie climbeth on trees, old buildings, and walls: the stalkes thereof are wooddy, and now and then so great as it seemes to become a tree; from which it sendeth a multitude of little boughes or branches every way, whereby as it were with armes it creepeth and wandereth far about: it also bringeth forth continually fine little roots, by which it fastneth it selfe and cleaveth wonderfull hard upon trees, and upon the smoothest stone walls: the leaves are smooth, shining especially on the upper side, cornered with sharpe pointed corners. The floures are very small and mossie; after which succeed bundles of black berries, every one having a small sharpe pointall.

Clymbing or berried Ivie

Ivie flourisheth in Autumne: the berries are ripe after the Winter Solstice.

The leaves laid in steepe in water for a day and a nights space, helpe sore and smarting waterish eies, if they be bathed and washed with the water wherein they have beene infused.

Misseltoe

Viscum or Misseltoe hath many slender branches spread overthwart one another, and wrapped or interlaced one within another, the barke wherof is of a light green or

Popinjay colour: the leaves of this branching excrescence be of a brown green colour: the floures be small and yellow: which beeing past, there appeare small clusters of white translucent berries, which are so cleare that a man may see thorow them, and are of a clammy or viscous moisture, whereof the best Bird-lime is made, far exceeding that which is made of the Holm or Holly bark: and

within this berry is a small blacke kernel or seed: this excrescence hath not any root, neither doth increase himselfe of his seed, as some have supposed; but it rather commeth of a certaine moisture and substance gathered together upon the boughs and joints of the trees, through the barke whereof this vaporous moisture proceeding, brings forth the Misseltoe. Many have diversly spoken hereof. Some of the Learned have set down, that it comes of the dung of the bird called a Thrush, who having fed of the seeds thereof, hath voided and left his dung upon the tree, whereof was ingendered

Misseltoe

this bery, a most fit matter to make lime of to intrap and catch birds withall.

Misseltoe groweth upon Okes and divers other trees almost every where.

Misseltoe is alwaies greene as well in winter as summer: the berries are ripe in Autumne, they remain all winter thorow, and are a food for divers birds, as Thrushes, Black-birds, & Ring-doves.

BLACKE HELLEBORE

This plant hath thicke and fat leaves of a deep green colour, the upper part whereof is somewhat bluntly nicked or toothed, having sundry divisions or cuts, in some leaves many, in others fewer. It beareth Rose-fashioned floures upon slender stems, growing immediatly out of the ground an handfull high, sometimes very white, and oftentimes mixed with a little shew of purple: which being vaded, there succeed small husks full of blacke seeds.

A purgation of Hellebor is good for mad and furious men, for melancholy, dull and heavie persons, and briefly for all those that are troubled with blacke choler, and molested with melancholy.

It is agreed among the later writers, that these plants are *Veratra nigra*: in English, blacke Hellebores: of divers, *Melampodium*, because it was first found by *Melampos*, who was first thought to purge therewith *Prætus* his mad daughters, and to restore them to health. *Dioscorides* writeth, that this man was a shepheard: others, a Soothsayer. In high Dutch it is called Christs herbe, and that because it floureth about the birth of our Lord Jesus Christ.

THE ARCHED INDIAN FIG TREE

This rare and admirable tree is very great, straight, and covered with a yellow barke tending to tawny: the boughes and branches are many, very long, tough, and flexible, growing very long in short space, as doe the twigs of Oziars, and those so long and weake, that the ends thereof hang downe and touch the ground, where they take root and grow in such sort, that those twigs become great trees: and these being growne up unto the like greatnesse, doe cast their branches or twiggy tendrels

283

unto the earth, where they likewise take hold and root; by meanes whereof it commeth to passe, that of one tree is made a great wood or desart of trees, which the Indians doe use for coverture against the extreme heat of the Sun, wherewith they are grievously vexed: some likewise use them for pleasure, cutting downe by a direct line a long walke, or as it were a vault, through the thickest part, from which also they cut certaine loope-holes or windowes in some places, to the end to receive thereby the fresh coole aire that entreth thereat, as also for light, that they may see their cattell that feed thereby, to avoid any danger that might happen unto them either by the enemy or wilde beasts: from which vault or close walke doth rebound such an admirable eccho or answering voice,

if one of them speake unto another aloud, that it doth resound or answer againe foure or five times, according to the height of the voice, to which it doth answer, and that so plainely, that it cannot be known from the voice it selfe: the first or mother of this wood or desart of trees is hard to bee knowne from the children, but by the greatnesse of the body, which three men can scarsely fathom about: upon the branches whereof grow leaves hard and wrinkled, in shape like those of the Quince tree, greene above, and of a whitish hoary colour underneath, whereupon the Elephants delight to feed: among which leaves

The arched Indian Fig tree

284

come forth the fruit, of the bignesse of a mans thumbe, in shape like a small Fig, but of a sanguine or bloudy colour, and of a sweet tast, but not so pleasant as the Figs of Spaine; notwithstanding they are good to be eaten, and withall very wholesome.

This wondrous tree groweth in divers places of the East-Indies, especially neere unto Goa, and also in Malaca: it is a stranger in most parts of the World. It keepeth his leaves greene Winter and Summer.

This tree is called of those that have travelled, *Ficus Indica*; the Indian Fig; and *Arbor Goa*, of the place where it groweth in greatest plenty: we may call it in English, the arched Fig tree.

THE GOOSE TREE, BARNACLE TREE, OR THE TREE BEARING GEESE

Having travelled from the Grasses growing in the bottome of the fenny waters, the Woods, and mountaines, even unto Libanus it selfe, wee are arrived at the end of our History; thinking it not impertinent to the conclusion of the same, to end with one of the marvels of this land (we may say of the World). The history whereof to set forth according to the worthinesse and raritie thereof, would not only require a large and peculiar volume, but also a deeper search into the bowels of Nature, than my intended purpose will suffer me to wade into, my sufficiencie also considered; leaving the History thereof rough hewen, unto some excellent man, learned in the secrets of nature, to be both fined and refined: in the meane space take it as it falleth out, the naked and bare truth, though unpolished. There are found in the North parts of Scotland and the Islands adjacent, called Orchades, certaine trees whereon do grow certaine shells of a white colour tending to russet, wherein are contained little living creatures: which shells in time of maturity doe open, and out of them grow those little living things, which falling into the water do become fowles, which we call Barnacles; in the North of England, brant Geese; and in Lancashire, tree Geese: but the other that do fall upon the land perish and come to nothing. Thus much by the writings of others, and also from the mouthes of people of those parts, which may very well accord with truth.

But what our eies have seene, and hands have touched we shall declare. There is a small Island in Lancashire called the Pile of Foulders, wherein are found the broken pieces of old and bruised ships, some whereof have beene cast thither by shipwracke, and also the trunks and bodies with the branches of old and rotten trees, cast up there likewise; whereon is found a certaine spume or froth that in time breedeth unto certaine shells, in shape like those of the Muskle, but sharper pointed, and of a whitish colour; wherein is contained a thing in forme like a lace of silke finely woven as it were together, of a whitish colour, one end whereof is fastned unto the inside of the shell, even as the fish of Oisters and Muskles are: the other end is made fast unto the belly of a rude masse or lumpe, which in time commeth to the shape and forme of a Bird: when it is perfectly formed the shell gapeth open, and the first thing that appeareth is the foresaid lace or string; next come the legs of the bird hanging out, and as it groweth greater it openeth the shell by degrees, til at length it is all come forth, and hangeth onely by the bill: in short space after it commeth to full maturitie, and falleth into the sea, where it gathereth feathers, and groweth to a fowle bigger than a Mallard, and lesser than a Goose, having blacke legs and bill or beake, and feathers blacke and white, spotted in such manner as is our Magpie, called in some places a Pie-Annet, which the people of Lancashire call by no other name than a tree Goose: which place aforesaid, and all those parts adjoyning do so much abound therewith, that one of the best is bought for three pence. For the truth hereof, if any doubt, may it please them to repair unto me, and I shall satisfie them by the testimonie of good witnesses.

Moreover, it should seeme that there is another sort hereof; the History of which is true, and of mine owne knowledge: for travelling upon the shore of our English coast betweene Dover and Rumney, I found the trunke of an old rotten tree, which (with some helpe that I procured by Fishermens wives that were there attending

The breed of Barnacles

their husbands returne from the sea) we drew out of the water upon dry land: upon this rotten tree I found growing many thousands of long crimson bladders, in shape like unto puddings newly filled, before they be sodden, which were very cleere and shining; at the nether end whereof did grow a shell fish, fashioned somewhat like a small Muskle, but much whiter, resembling a shell fish that groweth upon the rockes about Garnsey and Garsey, called a Lympit: many of these shells I brought with me to London, which after I had opened I found in them living things without forme or shape; in others which

287

were neerer come to ripenesse I found living things that were very naked, in shape like a Bird: in others, the Birds covered with soft downe, the shell halfe open, and the Bird ready to fall out, which no doubt were the Fowles called Barnacles. I dare not absolutely avouch every circumstance of the first part of this history, concerning the tree that beareth those buds aforesaid, but will leave it to a further consideration; howbeit, that which I have seene with mine eies, and handled with mine hands, I dare confidently avouch, and boldly put downe for verity. Now if any will object that this tree which I saw might be one of those before mentioned, which either by the waves of the sea or some violent wind had beene overturned as many other trees are; or that any trees falling into those seas about the Orchades, will of themselves beare the like Fowles, by reason of those seas and waters, these being so probable conjectures, and likely to be true, I may not without prejudice gainesay, or indeavour to confute.

And thus having through Gods assistance discoursed somewhat at large of Grasses, Herbes, Shrubs, Trees, and Mosses, and certaine Excrescences of the earth, with other things moe, incident to the historie thereof, we conclude and end our present Volume, with this wonder of England. For the which Gods Name be ever honored and praised.

FINIS

THE NOTES AND TABLES

NOTES

Botanical history compliments Gerard by giving the year 1597—
when his Herbal was published—or thereabouts as the date of the
introduction of many plants to Britain.   This date was convenient
when exact information was lacking and a plant had been described
by Gerard.   Examples are Margerome of Candy, Sugar Cane, an
Iris, some of the Lilies, and Crown Imperial.   Lettuce and Rosemary also were introduced in Gerard's time.

PAGE

4   The Spring Saffrons are the familiar *Crocus vernus* and
    C. luteus.
4   The Narcissi described are *Narcissus poeticus.*   The figure is of
    N. Jonquilla.
9   The first Anemone described is *A. coronaria,* of which the
    others are varieties.
20 The twin-like Cowslip is *Primula veris.*   Double Paigle and
    the Primrose with the greenish flowers are varieties of *P.
    vulgaris.*
23 *Anemone Pulsatilla* is a native of several parts of England; the
    white variety is rare.
24 The Sweet Johns and Sweet Williams are identified with
    Dianthus Carthusianorum.   The Pride of Austria is *D. superbus.*
26 The flowers which Gerard judged as too sweet have been taken
    for White Lilac, but the description favours *Philadelphus
    coronarius* (commonly called Syringa) which is the plant illus-
    trated.   " Blew Pipe " is the old name of *Syringa vulgaris,* the
    common Lilac.
42 The Docks first described are *Rumex Hydropapathum* and *R.
    obtusifolius.*   Bloudwort suggests *R. sanguineus. R. Patientia*
    came from Italy in Gerard's time.
53 Lily of the Valley still grows at Hampstead (in Ken Wood).

54 English Jacinth is Gerard's name for *Scilla nutans.* False bumbast Jacinth appears to be a species of *Hypoxis,* and Floure of Tygris the splendid *Tigridis Pavonia.*

57-60 White Lily and Lily of Constantinople are both *Lilium candidum.* Mountaine Lillies are *L. Martagon.* Persian Lilly is *Fritillaria persica,* and *Crowne Imperiall* (page 38) is *F. Imperialis.* Day Lillie is *Hemerocallis fulva.* *Fritillaria Meleagris* was not recognised as an English wild flower until the middle of the nineteenth century.

64 Gerard cultivated a dozen sorts of Iris. The figure is of *I. susiana,* introduced to Britain from the Levant 1596. Floure-de-luce of Dalmatia is *I. pallida.* The roots of *I. florentina* supply the orris root of the perfumers.

66 Gerard cultivated and described the Male, *Pæonia mascula* and the Female, *P. officinalis.* Stronger and weaker varieties of plants were often distinguished as male and female.

68 French Corne-flagge (*Gladiolus communis*) is a plant of central and southern Europe. In Britain it is found wild in the New Forest and the Isi᠍ of Wight.

71 In the expression "maketh a may gracious" the word "may" could mean either man or maid.

71 The first of these Bell-flowers is *Campanula persicifolia,* and the second *C. pyramidalis.*

76 Herb Two-pence (*Lysimachia Nummularia*) still has the popular name Money-wort, as well as Creeping Jenny.

77 White Hellebor is *Veratrum alba,* a sixteenth century introduction to England.

78 Although five British species of *Polygala* (or three with two varieties of one of them) are now generally recognised, at least three of the plants described are *P. vulgaris.* The illustration shows *Illecebrum verticillatum.*

82 Rough Bind-weed (*Smilax aspera*) is the Smilax of southern Europe. The smooth variety is, as Gerard says, not properly Smilax but *Convolvulus sepium,* and Scammonie is *C. scammonia,* imported to Britain from the Levant. *Smilax Sarsaparilla* was brought to Europe in the middle of the sixteenth century as a medicine of virtue.

87 The Swallow-wort described is *Vincetoxicum officinale.* The kind of Asclepias still bears the name of Æsculapius: *A. syriaca*

is the fragrant plant of Canadian woods, introduced to Britain in the seventeenth century.

90 The Limes described are *Tilia europæa*, supposed to be a hybrid, of garden origin.

94 It is remarkable that Gerard described no more than a dozen Roses. His White Rose is *Rosa arvensis*, the Red, *R. Gallica*, the Damaske, *R. provincialis*. The Great Holland Rose is *R. centifolia*, of which the Province Rose may have been a variety. The Muske Roses (page 99) are *R. moschata*, *R. Lutea* (illustrated) and *R. cinnamonea*.

104 The Campions described are both the exotic species, *Lychnis coronaria*.

108 Among the Foxgloves we recognise *Digitalis purpurea* and its white variety. It is remarkable that Gerard should accord the plants no place in medicine, considering that *Digitalis* ranks to-day as one of the most valuable of British "official herbs", of which about two dozen find a place in the *British Pharmacopœia*.

109, 114 Thora Valdensium is *Ranunculus Thora*. Yellow Wolfes-bane is *Aconitum Lycoctonum*, and Anthora is *Aconitum Anthora*.

115 The Sea Lavenders described are *Statice Limonium*, and *S. binervosa*, long regarded as a variety, but a distinct species in accordance with Gerard's decision.

116 *Aster Tripolium* is allied to the Michaelmas Daisy brought to England from North America by the elder John Tradescant, gardener to Charles I.

118 The two spurges described are *Euphorbia Paralias* and *E. Helioscopia*. The figure is of *E. amygdaloides*.

120 Gerard grouped various plants as Pennywort, of which there is one British species, *Cotyledon Umbilicus*. The second kind is *Sedum roseum*, and the third *Hydrocotyle vulgaris*.

124 Housleeke is *Sempervivum tectorum*.

126 Ladies Shoo is *Cypripedium Calceolus*.

136 Blacke Wortle is *Vaccinium Myrtillus* (Bilberry, Blaeberry, or Whortleberry) and red Whortle *V. Vitis-idæa* (Cowberry, or Red Whortleberry).

138 The red, white and green Strawberries are *Fragaria virginiana*, introduced from North America.

141 The first Reed is *Arundo Phragmites,* and the great sort is *A. Donax,* common as a screen in Italian gardens.

147 Some thirty Carnations, Gillofloures, Pinkes, Sweet-Johns, Sweet-Williams, and Wilde Williams are described in the Herbal. The "Gillofloure with yellow flours" is a variety of *Dianthus caryophyllus,* ancestor of the garden Carnation.

153 Mouse-eare Scorpion Grasse is an old name for the Forget-me-not. The first described is *Scorpiurus sulcatus.*

155 English Cudweed is *Gnaphalium sylvaticum,* one of the four species of the British genus of Everlasting Flowers. "Live-for-ever" is *G. margariticeum.* "Small" and "Wicked" Cud-weed are rightly named in the Herbal *Filago,* a genus of a few species formerly included with the other.

157 The Feverfews are *Chrysanthemum Parthenium,* the common British wild flower (perhaps not indigenous).

158 The female Mullein may have been either *Verbascum Blattaria* or *V. Lychnitis.*

159 The Purple Goat's-beard is Salsify (*Tragopogon porrifolius*) and is occasionally found established in southern England.

161 Oculus Christi is the British wild flower, Clary (*Salvia Verbenaca*); Purple Clarie is *S. Horminum.*

166 Dyer's Bugloss (*Anchusa tinctoria*) is a native of Italy. The *Echium* mentioned is Viper's Bugloss (*E. vulgare*).

167 Tarragon and *Draco herba* are old names for *Artemisia Dracunculus.*

168 Nasturtium (*Tropæolum*) is wrongly named as a Cress; it is a native of South America.

173 Three Teasels grow in Britain, of which *Dipsacus fullonum* is the famed Fuller's Teasel, possibly a cultivated variety. It is now known that the water-supply in Teasel cups is drawn from the soil.

174 Gerard cultivated and described several Rues not represented in Britain, as Garden Rue (*Ruta graveolens*) and wild Rue (*R. montana.*)

178 Skirrets, the Water Parsnip (*Sium Sisarum*) was brought to England from China; though rare, it may linger in cottage gardens.

191 Thorn-Apple (*Datura Stramonium*) is an instance of a South American plant which in a short time made itself more or less at

home in southern England, the seeds having been introduced from Constantinople (or, according to some, from Italy or Spain) and dispersed through the land by Gerard. In Virginia it is named Fire-weed, from springing up after fires. The plant first described is *Datura Metel*; these names are from the Arabic, the specific one expressing the narcotic effect of the plant.

198, 199, 200 Common Henbane (*Hyoscyamus niger*) is the common British weed, also of Europe and Western Asia. Yellow Henbane is *Nicotiana rustica*. Tobacco plants, said to have been first brought from America in 1570, were named *Nicotiana* in honour of John Nicot, French Ambassador in Portugal, who procured the seeds.

208 Floure-gentle excited delight when introduced to Britain from the East Indies, as "very brave to look upon". The plant first described is identified with *Celosia cristata*, and the Floramor is *Amaranthus tricolor*.

216 Great Blew-Bottle is *Centaurea montana*, introduced to Britain from central and southern Europe. Gerard also cultivated *C. Cyanus*.

222 The Woundwort is *Stachys palustris*, so named by Gerard's first editor, Johnson.

225 Early herbalists termed the Scarlet Pimpernel (*Anagallis arvensis*) the Male, and the blue variety (*A. cœrulea*) the Female. The plant figured in *Anagallis monelli*.

228 Bastard Margerome of Candy (*Origanum creticum*) was introduced to Britain from Southern Europe about 1597.

231 *Verbena officinalis* is the one British representative of the genus, so that it is doubtful if Gerard found Creeping Vervaine growing wild.

243 The great Maple is *Acer pseudo-platanus*, naturalised in Britain.

255 The names White Satin, *Lunaria*, and Pricksong-wort are allusions to the seed-vessels. "Pricks" were notes in written music, and a sheet of music in Gerard's day was a pricksong.

258 The Meadow Saffrons described are all *Colchicum autumnalis*. *Hermodactylus* was a name given to an Iris.

265 *Zea Mays* was introduced to Britain from North America, 1562.

267-8 Gerard published the first picture of the so-called Virginian Potato (not a native of Virginia) supposed to have been brought to Britain from that country by colonists sent out by Sir Walter Raleigh. The name distinguished this from "Battatas", or Sweet Potatoes, used in England as a delicacy before the introduction of the other, and confusingly called by Gerard the Common Potato. In Gerard's portrait in the Herbal he is holding a branch of the Potato.

274 The name "Traveller's Joy" was one of Gerard's happiest inventions. Though he objected to the name *Vitis alba* it remains that of the one British species. *Viorna* is a North American species.

281 Mistletoe, contrary to Gerard's (and his artist's) opinion, grows very rarely on Oaks. It is surprising that after mentioning the seed he should have denied that it increases by seed.

285 The myth of the Barnacle Tree goes back to remote ages. In 1677 a paper was read on the subject before the Royal Society; Gerard was in good company (including that of the herbalist William Turner) in giving credence to the legend.

TABLE OF SUNDRY VERTUES

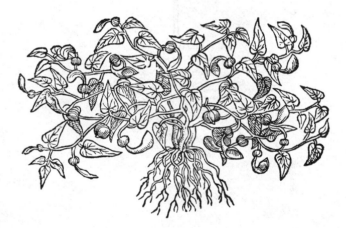

ALPHABETICAL TABLE OF PLANTS

A CATALOGUE OF SELECTED DOVER BOOKS
IN ALL FIELDS OF INTEREST

A CATALOGUE OF SELECTED DOVER BOOKS
IN ALL FIELDS OF INTEREST

WHAT IS SCIENCE?, *N. Campbell*
The role of experiment and measurement, the function of mathematics, the nature of scientific laws, the difference between laws and theories, the limitations of science, and many similarly provocative topics are treated clearly and without technicalities by an eminent scientist. "Still an excellent introduction to scientific philosophy," H. Margenau in *Physics Today.* "A first-rate primer . . . deserves a wide audience," *Scientific American.* 192pp. 5⅜ x 8.
S43　Paperbound $1.25

THE NATURE OF LIGHT AND COLOUR IN THE OPEN AIR, *M. Minnaert*
Why are shadows sometimes blue, sometimes green, or other colors depending on the light and surroundings? What causes mirages? Why do multiple suns and moons appear in the sky? Professor Minnaert explains these unusual phenomena and hundreds of others in simple, easy-to-understand terms based on optical laws and the properties of light and color. No mathematics is required but artists, scientists, students, and everyone fascinated by these "tricks" of nature will find thousands of useful and amazing pieces of information. Hundreds of observational experiments are suggested which require no special equipment. 200 illustrations; 42 photos. xvi + 362pp. 5⅜ x 8.
T196　Paperbound $2.00

THE STRANGE STORY OF THE QUANTUM, AN ACCOUNT FOR THE GENERAL READER OF THE GROWTH OF IDEAS UNDERLYING OUR PRESENT ATOMIC KNOWLEDGE, *B. Hoffmann*
Presents lucidly and expertly, with barest amount of mathematics, the problems and theories which led to modern quantum physics. Dr. Hoffmann begins with the closing years of the 19th century, when certain trifling discrepancies were noticed, and with illuminating analogies and examples takes you through the brilliant concepts of Planck, Einstein, Pauli, Broglie, Bohr, Schroedinger, Heisenberg, Dirac, Sommerfeld, Feynman, etc. This edition includes a new, long postscript carrying the story through 1958. "Of the books attempting an account of the history and contents of our modern atomic physics which have come to my attention, this is the best," H. Margenau, Yale University, in *American Journal of Physics.* 32 tables and line illustrations. Index. 275pp. 5⅜ x 8.
T518　Paperbound $2.00

GREAT IDEAS OF MODERN MATHEMATICS: THEIR NATURE AND USE, *Jagjit Singh*
Reader with only high school math will understand main mathematical ideas of modern physics, astronomy, genetics, psychology, evolution, etc. better than many who use them as tools, but comprehend little of their basic structure. Author uses his wide knowledge of non-mathematical fields in brilliant exposition of differential equations, matrices, group theory, logic, statistics, problems of mathematical foundations, imaginary numbers, vectors, etc. Original publication. 2 appendixes. 2 indexes. 65 ills. 322pp. 5⅜ x 8.
T587　Paperbound $2.25

THE MUSIC OF THE SPHERES: THE MATERIAL UNIVERSE — FROM ATOM TO QUASAR, SIMPLY EXPLAINED, *Guy Murchie*
Vast compendium of fact, modern concept and theory, observed and calculated data, historical background guides intelligent layman through the material universe. Brilliant exposition of earth's construction, explanations for moon's craters, atmospheric components of Venus and Mars (with data from recent fly-by's), sun spots, sequences of star birth and death, neighboring galaxies, contributions of Galileo, Tycho Brahe, Kepler, etc.; and (Vol. 2) construction of the atom (describing newly discovered sigma and xi subatomic particles), theories of sound, color and light, space and time, including relativity theory, quantum theory, wave theory, probability theory, work of Newton, Maxwell, Faraday, Einstein, de Broglie, etc. "Best presentation yet offered to the intelligent general reader," *Saturday Review*. Revised (1967). Index. 319 illustrations by the author. Total of xx + 644pp. 5⅜ x 8½.
T1809, T1810   Two volume set, paperbound $4.00

FOUR LECTURES ON RELATIVITY AND SPACE, *Charles Proteus Steinmetz*
Lecture series, given by great mathematician and electrical engineer, generally considered one of the best popular-level expositions of special and general relativity theories and related questions. Steinmetz translates complex mathematical reasoning into language accessible to laymen through analogy, example and comparison. Among topics covered are relativity of motion, location, time; of mass; acceleration; 4-dimensional time-space; geometry of the gravitational field; curvature and bending of space; non-Euclidean geometry. Index. 40 illustrations. x + 142pp. 5⅜ x 8½.   S1771   Paperbound $1.35

HOW TO KNOW THE WILD FLOWERS, *Mrs. William Starr Dana*
Classic nature book that has introduced thousands to wonders of American wild flowers. Color-season principle of organization is easy to use, even by those with no botanical training, and the genial, refreshing discussions of history, folklore, uses of over 1,000 native and escape flowers, foliage plants are informative as well as fun to read. Over 170 full-page plates, collected from several editions, may be colored in to make permanent records of finds. Revised to conform with 1950 edition of Gray's Manual of Botany. xlii + 438pp. 5⅜ x 8½.   T332   Paperbound $2.25

MANUAL OF THE TREES OF NORTH AMERICA, *Charles Sprague Sargent*
Still unsurpassed as most comprehensive, reliable study of North American tree characteristics, precise locations and distribution. By dean of American dendrologists. Every tree native to U.S., Canada, Alaska; 185 genera, 717 species, described in detail—leaves, flowers, fruit, winterbuds, bark, wood, growth habits, etc. plus discussion of varieties and local variants, immaturity variations. Over 100 keys, including unusual 11-page analytical key to genera, aid in identification. 783 clear illustrations of flowers, fruit, leaves. An unmatched permanent reference work for all nature lovers. Second enlarged (1926) edition. Synopsis of families. Analytical key to genera. Glossary of technical terms. Index. 783 illustrations, 1 map. Total of 982pp. 5⅜ x 8.
T277, T278   Two volume set, paperbound $6.00

IT'S FUN TO MAKE THINGS FROM SCRAP MATERIALS,
Evelyn Glantz Hershoff
What use are empty spools, tin cans, bottle tops? What can be made from rubber bands, clothes pins, paper clips, and buttons? This book provides simply worded instructions and large diagrams showing you how to make cookie cutters, toy trucks, paper turkeys, Halloween masks, telephone sets, aprons, linoleum block- and spatter prints — in all 399 projects! Many are easy enough for young children to figure out for themselves; some challenging enough to entertain adults; all are remarkably ingenious ways to make things from materials that cost pennies or less! Formerly "Scrap Fun for Everyone." Index. 214 illustrations. 373pp. 5⅜ x 8½.   T1251 Paperbound $1.75

SYMBOLIC LOGIC and THE GAME OF LOGIC, *Lewis Carroll*
"Symbolic Logic" is not concerned with modern symbolic logic, but is instead a collection of over 380 problems posed with charm and imagination, using the syllogism and a fascinating diagrammatic method of drawing conclusions. In "The Game of Logic" Carroll's whimsical imagination devises a logical game played with 2 diagrams and counters (included) to manipulate hundreds of tricky syllogisms. The final section, "Hit or Miss" is a lagniappe of 101 additional puzzles in the delightful Carroll manner. Until this reprint edition, both of these books were rarities costing up to $15 each. Symbolic Logic: Index. xxxi + 199pp. The Game of Logic: 96pp. 2 vols. bound as one. 5⅜ x 8.
T492 Paperbound $2.00

MATHEMATICAL PUZZLES OF SAM LOYD, PART I
selected and edited by M. Gardner
Choice puzzles by the greatest American puzzle creator and innovator. Selected from his famous collection, "Cyclopedia of Puzzles," they retain the unique style and historical flavor of the originals. There are posers based on arithmetic, algebra, probability, game theory, route tracing, topology, counter and sliding block, operations research, geometrical dissection. Includes the famous "14-15" puzzle which was a national craze, and his "Horse of a Different Color" which sold millions of copies. 117 of his most ingenious puzzles in all. 120 line drawings and diagrams. Solutions. Selected references. xx + 167pp. 5⅜ x 8.
T498 Paperbound $1.25

STRING FIGURES AND HOW TO MAKE THEM, *Caroline Furness Jayne*
107 string figures plus variations selected from the best primitive and modern examples developed by Navajo, Apache, pygmies of Africa, Eskimo, in Europe, Australia, China, etc. The most readily understandable, easy-to-follow book in English on perennially popular recreation. Crystal-clear exposition; step-by-step diagrams. Everyone from kindergarten children to adults looking for unusual diversion will be endlessly amused. Index. Bibliography. Introduction by A. C. Haddon. 17 full-page plates, 960 illustrations. xxiii + 401pp. 5⅜ x 8½.
T152 Paperbound $2.25

PAPER FOLDING FOR BEGINNERS, *W. D. Murray and F. J. Rigney*
A delightful introduction to the varied and entertaining Japanese art of origami (paper folding), with a full, crystal-clear text that anticipates every difficulty; over 275 clearly labeled diagrams of all important stages in creation. You get results at each stage, since complex figures are logically developed from simpler ones. 43 different pieces are explained: sailboats, frogs, roosters, etc. 6 photographic plates. 279 diagrams. 95pp. 5⅝ x 8⅜.
T713 Paperbound $1.00

PRINCIPLES OF ART HISTORY,
H. Wölfflin
Analyzing such terms as "baroque," "classic," "neoclassic," "primitive," "picturesque," and 164 different works by artists like Botticelli, van Cleve, Dürer, Hobbema, Holbein, Hals, Rembrandt, Titian, Brueghel, Vermeer, and many others, the author establishes the classifications of art history and style on a firm, concrete basis. This classic of art criticism shows what really occurred between the 14th-century primitives and the sophistication of the 18th century in terms of basic attitudes and philosophies. "A remarkable lesson in the art of seeing," *Sat. Rev. of Literature.* Translated from the 7th German edition. 150 illustrations. 254pp. 6⅛ x 9¼. T276 Paperbound $2.00

PRIMITIVE ART,
Franz Boas
This authoritative and exhaustive work by a great American anthropologist covers the entire gamut of primitive art. Pottery, leatherwork, metal work, stone work, wood, basketry, are treated in detail. Theories of primitive art, historical depth in art history, technical virtuosity, unconscious levels of patterning, symbolism, styles, literature, music, dance, etc. A must book for the interested layman, the anthropologist, artist, handicrafter (hundreds of unusual motifs), and the historian. Over 900 illustrations (50 ceramic vessels, 12 totem poles, etc.). 376pp. 5⅜ x 8. T25 Paperbound $2.50

THE GENTLEMAN AND CABINET MAKER'S DIRECTOR,
Thomas Chippendale
A reprint of the 1762 catalogue of furniture designs that went on to influence generations of English and Colonial and Early Republic American furniture makers. The 200 plates, most of them full-page sized, show Chippendale's designs for French (Louis XV), Gothic, and Chinese-manner chairs, sofas, canopy and dome beds, cornices, chamber organs, cabinets, shaving tables, commodes, picture frames, frets, candle stands, chimney pieces, decorations, etc. The drawings are all elegant and highly detailed; many include construction diagrams and elevations. A supplement of 24 photographs shows surviving pieces of original and Chippendale-style pieces of furniture. Brief biography of Chippendale by N. I. Bienenstock, editor of *Furniture World.* Reproduced from the 1762 edition. 200 plates, plus 19 photographic plates. vi + 249pp. 9⅛ x 12¼. T1601 Paperbound $3.50

AMERICAN ANTIQUE FURNITURE: A BOOK FOR AMATEURS,
Edgar G. Miller, Jr.
Standard introduction and practical guide to identification of valuable American antique furniture. 2115 illustrations, mostly photographs taken by the author in 148 private homes, are arranged in chronological order in extensive chapters on chairs, sofas, chests, desks, bedsteads, mirrors, tables, clocks, and other articles. Focus is on furniture accessible to the collector, including simpler pieces and a larger than usual coverage of Empire style. Introductory chapters identify structural elements, characteristics of various styles, how to avoid fakes, etc. "We are frequently asked to name some book on American furniture that will meet the requirements of the novice collector, the beginning dealer, and . . . the general public. . . . We believe Mr. Miller's two volumes more completely satisfy this specification than any other work," *Antiques.* Appendix. Index. Total of vi + 1106pp. 7⅞ x 10¾.
T1599, T1600 Two volume set, paperbound $7.50

THE BAD CHILD'S BOOK OF BEASTS, MORE BEASTS FOR WORSE CHILDREN, and A MORAL ALPHABET, *H. Belloc*
Hardly and anthology of humorous verse has appeared in the last 50 years without at least a couple of these famous nonsense verses. But one must see the entire volumes — with all the delightful original illustrations by Sir Basil Blackwood — to appreciate fully Belloc's charming and witty verses that play so subacidly on the platitudes of life and morals that beset his day — and ours. A great humor classic. Three books in one. Total of 157pp. 5⅜ x 8.
T749 Paperbound $1.00

THE DEVIL'S DICTIONARY, *Ambrose Bierce*
Sardonic and irreverent barbs puncturing the pomposities and absurdities of American politics, business, religion, literature, and arts, by the country's greatest satirist in the classic tradition. Epigrammatic as Shaw, piercing as Swift, American as Mark Twain, Will Rogers, and Fred Allen, Bierce will always remain the favorite of a small coterie of enthusiasts, and of writers and speakers whom he supplies with "some of the most gorgeous witticisms of the English language" (H. L. Mencken). Over 1000 entries in alphabetical order. 144pp. 5⅜ x 8. T487 Paperbound $1.00

THE COMPLETE NONSENSE OF EDWARD LEAR.
This is the only complete edition of this master of gentle madness available at a popular price. *A Book of Nonsense, Nonsense Songs, More Nonsense Songs and Stories* in their entirety with all the old favorites that have delighted children and adults for years. The Dong With A Luminous Nose, The Jumblies, The Owl and the Pussycat, and hundreds of other bits of wonderful nonsense: 214 limericks, 3 sets of Nonsense Botany, 5 Nonsense Alphabets, 546 drawings by Lear himself, and much more. 320pp. 5⅜ x 8. T167 Paperbound $1.75

THE WIT AND HUMOR OF OSCAR WILDE, *ed. by Alvin Redman*
Wilde at his most brilliant, in 1000 epigrams exposing weaknesses and hypocrisies of "civilized" society. Divided into 49 categories—sin, wealth, women, America, etc.—to aid writers, speakers. Includes excerpts from his trials, books, plays, criticism. Formerly "The Epigrams of Oscar Wilde." Introduction by Vyvyan Holland, Wilde's only living son. Introductory essay by editor. 260pp. 5⅜ x 8. T602 Paperbound $1.50

A CHILD'S PRIMER OF NATURAL HISTORY, *Oliver Herford*
Scarcely an anthology of whimsy and humor has appeared in the last 50 years without a contribution from Oliver Herford. Yet the works from which these examples are drawn have been almost impossible to obtain! Here at last are Herford's improbable definitions of a menagerie of familiar and weird animals, each verse illustrated by the author's own drawings. 24 drawings in 2 colors; 24 additional drawings. vii + 95pp. 6½ x 6. T1647 Paperbound $1.00

THE BROWNIES: THEIR BOOK, *Palmer Cox*
The book that made the Brownies a household word. Generations of readers have enjoyed the antics, predicaments and adventures of these jovial sprites, who emerge from the forest at night to play or to come to the aid of a deserving human. Delightful illustrations by the author decorate nearly every page. 24 short verse tales with 266 illustrations. 155pp. 6⅝ x 9¼.
T1265 Paperbound $1.50

THE PRINCIPLES OF PSYCHOLOGY,
William James
The full long-course, unabridged, of one of the great classics of Western literature and science. Wonderfully lucid descriptions of human mental activity, the stream of thought, consciousness, time perception, memory, imagination, emotions, reason, abnormal phenomena, and similar topics. Original contributions are integrated with the work of such men as Berkeley, Binet, Mills, Darwin, Hume, Kant, Royce, Schopenhauer, Spinoza, Locke, Descartes, Galton, Wundt, Lotze, Herbart, Fechner, and scores of others. All contrasting interpretations of mental phenomena are examined in detail—introspective analysis, philosophical interpretation, and experimental research. "A classic," *Journal of Consulting Psychology*. "The main lines are as valid as ever," *Psychoanalytical Quarterly*. "Standard reading . . . a classic of interpretation," *Psychiatric Quarterly*. 94 illustrations. 1408pp. 5⅜ x 8.
T381, T382   Two volume set, paperbound $6.00

VISUAL ILLUSIONS: THEIR CAUSES, CHARACTERISTICS AND APPLICATIONS,
M. Luckiesh
"Seeing is deceiving," asserts the author of this introduction to virtually every type of optical illusion known. The text both describes and explains the principles involved in color illusions, figure-ground, distance illusions, etc. 100 photographs, drawings and diagrams prove how easy it is to fool the sense: circles that aren't round, parallel lines that seem to bend, stationary figures that seem to move as you stare at them — illustration after illustration strains our credulity at what we see. Fascinating book from many points of view, from applications for artists, in camouflage, etc. to the psychology of vision. New introduction by William Ittleson, Dept. of Psychology, Queens College. Index. Bibliography. xxi + 252pp. 5⅜ x 8½.   T1530   Paperbound $1.50

FADS AND FALLACIES IN THE NAME OF SCIENCE,
Martin Gardner
This is the standard account of various cults, quack systems, and delusions which have masqueraded as science: hollow earth fanatics. Reich and orgone sex energy, dianetics, Atlantis, multiple moons, Forteanism, flying saucers, medical fallacies like iridiagnosis, zone therapy, etc. A new chapter has been added on Bridey Murphy, psionics, and other recent manifestations in this field. This is a fair, reasoned appraisal of eccentric theory which provides excellent inoculation against cleverly masked nonsense. "Should be read by everyone, scientist and non-scientist alike," R. T. Birge, Prof. Emeritus of Physics, Univ. of California; Former President, American Physical Society. Index. x + 365pp. 5⅜ x 8.   T394   Paperbound $2.00

ILLUSIONS AND DELUSIONS OF THE SUPERNATURAL AND THE OCCULT,
D. H. Rawcliffe
Holds up to rational examination hundreds of persistent delusions including crystal gazing, automatic writing, table turning, mediumistic trances, mental healing, stigmata, lycanthropy, live burial, the Indian Rope Trick, spiritualism, dowsing, telepathy, clairvoyance, ghosts, ESP, etc. The author explains and exposes the mental and physical deceptions involved, making this not only an exposé of supernatural phenomena, but a valuable exposition of characteristic types of abnormal psychology. Originally titled "The Psychology of the Occult." 14 illustrations. Index. 551pp. 5⅜ x 8. T503 Paperbound $2.75

FAIRY TALE COLLECTIONS, *edited by Andrew Lang*
Andrew Lang's fairy tale collections make up the richest shelf-full of traditional children's stories anywhere available. Lang supervised the translation of stories from all over the world—familiar European tales collected by Grimm, animal stories from Negro Africa, myths of primitive Australia, stories from Russia, Hungary, Iceland, Japan, and many other countries. Lang's selection of translations are unusually high; many authorities consider that the most familiar tales find their best versions in these volumes. All collections are richly decorated and illustrated by H. J. Ford and other artists.

THE BLUE FAIRY BOOK. 37 stories. 138 illustrations. ix + 390pp. 5⅜ x 8½.
T1437 Paperbound $1.95

THE GREEN FAIRY BOOK. 42 stories. 100 illustrations. xiii + 366pp. 5⅜ x 8½.
T1439 Paperbound $1.75

THE BROWN FAIRY BOOK. 32 stories. 50 illustrations, 8 in color. xii + 350pp. 5⅜ x 8½.
T1438 Paperbound $1.95

THE BEST TALES OF HOFFMANN, *edited by E. F. Bleiler*
10 stories by E. T. A. Hoffmann, one of the greatest of all writers of fantasy. The tales include "The Golden Flower Pot," "Automata," "A New Year's Eve Adventure," "Nutcracker and the King of Mice," "Sand-Man," and others. Vigorous characterizations of highly eccentric personalities, remarkably imaginative situations, and intensely fast pacing has made these tales popular all over the world for 150 years. Editor's introduction. 7 drawings by Hoffmann. xxxiii + 419pp. 5⅜ x 8½.
T1793 Paperbound $2.25

GHOST AND HORROR STORIES OF AMBROSE BIERCE, *edited by E. F. Bleiler*
Morbid, eerie, horrifying tales of possessed poets, shabby aristocrats, revived corpses, and haunted malefactors. Widely acknowledged as the best of their kind between Poe and the moderns, reflecting their author's inner torment and bitter view of life. Includes "Damned Thing," "The Middle Toe of the Right Foot," "The Eyes of the Panther," "Visions of the Night," "Moxon's Master," and over a dozen others. Editor's introduction. xxii + 199pp. 5⅜ x 8½.
T767 Paperbound $1.50

THREE GOTHIC NOVELS, *edited by E. F. Bleiler*
Originators of the still popular Gothic novel form, influential in ushering in early 19th-century Romanticism. Horace Walpole's *Castle of Otranto*, William Beckford's *Vathek*, John Polidori's *The Vampyre*, and a *Fragment* by Lord Byron are enjoyable as exciting reading or as documents in the history of English literature. Editor's introduction. xi + 291pp. 5⅜ x 8½.
T1232 Paperbound $2.00

BEST GHOST STORIES OF LEFANU, *edited by E. F. Bleiler*
Though admired by such critics as V. S. Pritchett, Charles Dickens and Henry James, ghost stories by the Irish novelist Joseph Sheridan LeFanu have never become as widely known as his detective fiction. About half of the 16 stories in this collection have never before been available in America. Collection includes "Carmilla" (perhaps the best vampire story ever written), "The Haunted Baronet," "The Fortunes of Sir Robert Ardagh," and the classic "Green Tea." Editor's introduction. 7 contemporary illustrations. Portrait of LeFanu. xii + 467pp. 5⅜ x 8.
T415 Paperbound $2.50

EASY-TO-DO ENTERTAINMENTS AND DIVERSIONS WITH COINS, CARDS, STRING, PAPER AND MATCHES, *R. M. Abraham*
Over 300 tricks, games and puzzles will provide young readers with absorbing fun. Sections on card games; paper-folding; tricks with coins, matches and pieces of string; games for the agile; toy-making from common household objects; mathematical recreations; and 50 miscellaneous pastimes. Anyone in charge of groups of youngsters, including hard-pressed parents, and in need of suggestions on how to keep children sensibly amused and quietly content will find this book indispensable. Clear, simple text, copious number of delightful line drawings and illustrative diagrams. Originally titled "Winter Nights' Entertainments." Introduction by Lord Baden Powell. 329 illustrations. v + 186pp. 5⅜ x 8½. T921 Paperbound $1.00

AN INTRODUCTION TO CHESS MOVES AND TACTICS SIMPLY EXPLAINED, *Leonard Barden*
Beginner's introduction to the royal game. Names, possible moves of the pieces, definitions of essential terms, how games are won, etc. explained in 30-odd pages. With this background you'll be able to sit right down and play. Balance of book teaches strategy — openings, middle game, typical endgame play, and suggestions for improving your game. A sample game is fully analyzed. True middle-level introduction, teaching you all the essentials without oversimplifying or losing you in a maze of detail. 58 figures. 102pp. 5⅜ x 8½. T1210 Paperbound $1.25

LASKER'S MANUAL OF CHESS, *Dr. Emanuel Lasker*
Probably the greatest chess player of modern times, Dr. Emanuel Lasker held the world championship 28 years, independent of passing schools or fashions. This unmatched study of the game, chiefly for intermediate to skilled players, analyzes basic methods, combinations, position play, the aesthetics of chess, dozens of different openings, etc., with constant reference to great modern games. Contains a brilliant exposition of Steinitz's important theories. Introduction by Fred Reinfeld. Tables of Lasker's tournament record. 3 indices. 308 diagrams. 1 photograph. xxx + 349pp. 5⅜ x 8. T640 Paperbound $2.50

COMBINATIONS: THE HEART OF CHESS, *Irving Chernev*
Step-by-step from simple combinations to complex, this book, by a well-known chess writer, shows you the intricacies of pins, counter-pins, knight forks, and smothered mates. Other chapters show alternate lines of play to those taken in actual championship games; boomerang combinations; classic examples of brilliant combination play by Nimzovich, Rubinstein, Tarrasch, Botvinnik, Alekhine and Capablanca. Index. 356 diagrams. ix + 245pp. 5⅜ x 8½. T1744 Paperbound $2.00

HOW TO SOLVE CHESS PROBLEMS, *K. S. Howard*
Full of practical suggestions for the fan or the beginner — who knows only the moves of the chessmen. Contains preliminary section and 58 two-move, 46 three-move, and 8 four-move problems composed by 27 outstanding American problem creators in the last 30 years. Explanation of all terms and exhaustive index. "Just what is wanted for the student," Brian Harley. 112 problems, solutions. vi + 171pp. 5⅜ x 8. T748 Paperbound $1.35

SOCIAL THOUGHT FROM LORE TO SCIENCE,
H. E. Barnes and H. Becker
An immense survey of sociological thought and ways of viewing, studying, planning, and reforming society from earliest times to the present. Includes thought on society of preliterate peoples, ancient non-Western cultures, and every great movement in Europe, America, and modern Japan. Analyzes hundreds of great thinkers: Plato, Augustine, Bodin, Vico, Montesquieu, Herder, Comte, Marx, etc. Weighs the contributions of utopians, sophists, fascists and communists; economists, jurists, philosophers, ecclesiastics, and every 19th and 20th century school of scientific sociology, anthropology, and social psychology throughout the world. Combines topical, chronological, and regional approaches, treating the evolution of social thought as a process rather than as a series of mere topics. "Impressive accuracy, competence, and discrimination . . . easily the best single survey," *Nation*. Thoroughly revised, with new material up to 1960. 2 indexes. Over 2200 bibliographical notes. Three volume set. Total of 1586pp. 5⅜ x 8.

T901, T902, T903 Three volume set, paperbound $9.00

A HISTORY OF HISTORICAL WRITING, *Harry Elmer Barnes*
Virtually the only adequate survey of the whole course of historical writing in a single volume. Surveys developments from the beginnings of historiography in the ancient Near East and the Classical World, up through the Cold War. Covers major historians in detail, shows interrelationship with cultural background, makes clear individual contributions, evaluates and estimates importance; also enormously rich upon minor authors and thinkers who are usually passed over. Packed with scholarship and learning, clear, easily written. Indispensable to every student of history. Revised and enlarged up to 1961. Index and bibliography. xv + 442pp. 5⅜ x 8½.

T104 Paperbound $2.50

JOHANN SEBASTIAN BACH, *Philipp Spitta*
The complete and unabridged text of the definitive study of Bach. Written some 70 years ago, it is still unsurpassed for its coverage of nearly all aspects of Bach's life and work. There could hardly be a finer non-technical introduction to Bach's music than the detailed, lucid analyses which Spitta provides for hundreds of individual pieces. 26 solid pages are devoted to the B minor mass, for example, and 30 pages to the glorious St. Matthew Passion. This monumental set also includes a major analysis of the music of the 18th century: Buxtehude, Pachelbel, etc. "Unchallenged as the last word on one of the supreme geniuses of music," John Barkham, *Saturday Review Syndicate*. Total of 1819pp. Heavy cloth binding. 5⅜ x 8.

T252 Two volume set, clothbound $15.00

BEETHOVEN AND HIS NINE SYMPHONIES, *George Grove*
In this modern middle-level classic of musicology Grove not only analyzes all nine of Beethoven's symphonies very thoroughly in terms of their musical structure, but also discusses the circumstances under which they were written, Beethoven's stylistic development, and much other background material. This is an extremely rich book, yet very easily followed; it is highly recommended to anyone seriously interested in music. Over 250 musical passages. Index. viii + 407pp. 5⅜ x 8.

T334 Paperbound $2.25

THREE SCIENCE FICTION NOVELS,
John Taine
Acknowledged by many as the best SF writer of the 1920's, Taine (under the name Eric Temple Bell) was also a Professor of Mathematics of considerable renown. Reprinted here are *The Time Stream*, generally considered Taine's best, *The Greatest Game*, a biological-fiction novel, and *The Purple Sapphire*, involving a supercivilization of the past. Taine's stories tie fantastic narratives to frameworks of original and logical scientific concepts. Speculation is often profound on such questions as the nature of time, concept of entropy, cyclical universes, etc. 4 contemporary illustrations. v + 532pp. 5⅜ x 8⅜.
T1180 Paperbound $2.00

SEVEN SCIENCE FICTION NOVELS,
H. G. Wells
Full unabridged texts of 7 science-fiction novels of the master. Ranging from biology, physics, chemistry, astronomy, to sociology and other studies, Mr. Wells extrapolates whole worlds of strange and intriguing character. "One will have to go far to match this for entertainment, excitement, and sheer pleasure . . ."*New York Times*. Contents: The Time Machine, The Island of Dr. Moreau, The First Men in the Moon, The Invisible Man, The War of the Worlds, The Food of the Gods, In The Days of the Comet. 1015pp. 5⅜ x 8.
T264 Clothbound $5.00

28 SCIENCE FICTION STORIES OF H. G. WELLS.
Two full, unabridged novels, *Men Like Gods* and *Star Begotten*, plus 26 short stories by the master science-fiction writer of all time! Stories of space, time, invention, exploration, futuristic adventure. Partial contents: *The Country of the Blind, In the Abyss, The Crystal Egg, The Man Who Could Work Miracles, A Story of Days to Come, The Empire of the Ants, The Magic Shop, The Valley of the Spiders, A Story of the Stone Age, Under the Knife, Sea Raiders*, etc. An indispensable collection for the library of anyone interested in science fiction adventure. 928pp. 5⅜ x 8. T265 Clothbound $5.00

THREE MARTIAN NOVELS,
Edgar Rice Burroughs
Complete, unabridged reprinting, in one volume, of Thuvia, Maid of Mars; Chessmen of Mars; The Master Mind of Mars. Hours of science-fiction adventure by a modern master storyteller. Reset in large clear type for easy reading. 16 illustrations by J. Allen St. John. vi + 490pp. 5⅜ x 8½.
T39 Paperbound $2.50

AN INTELLECTUAL AND CULTURAL HISTORY OF THE WESTERN WORLD,
Harry Elmer Barnes
Monumental 3-volume survey of intellectual development of Europe from primitive cultures to the present day. Every significant product of human intellect traced through history: art, literature, mathematics, physical sciences, medicine, music, technology, social sciences, religions, jurisprudence, education, etc. Presentation is lucid and specific, analyzing in detail specific discoveries, theories, literary works, and so on. Revised (1965) by recognized scholars in specialized fields under the direction of Prof. Barnes. Revised bibliography. Indexes. 24 illustrations. Total of xxix + 1318pp.
T1275, T1276, T1277 Three volume set, paperbound $7.50

HEAR ME TALKIN' TO YA, *edited by Nat Shapiro and Nat Hentoff*
In their own words, Louis Armstrong, King Oliver, Fletcher Henderson, Bunk
Johnson, Bix Beiderbecke, Billy Holiday, Fats Waller, Jelly Roll Morton,
Duke Ellington, and many others comment on the origins of jazz in New
Orleans and its growth in Chicago's South Side, Kansas City's jam sessions,
Depression Harlem, and the modernism of the West Coast schools. Taken
from taped conversations, letters, magazine articles, other first-hand sources.
Editors' introduction. xvi + 429pp. 5⅜ x 8½. T1726 Paperbound $2.00

THE JOURNAL OF HENRY D. THOREAU
A 25-year record by the great American observer and critic, as complete a
record of a great man's inner life as is anywhere available. Thoreau's Journals
served him as raw material for his formal pieces, as a place where he could
develop his ideas, as an outlet for his interests in wild life and plants, in
writing as an art, in classics of literature, Walt Whitman and other contemporaries, in politics, slavery, individual's relation to the State, etc. The
Journals present a portrait of a remarkable man, and are an observant social
history. Unabridged republication of 1906 edition, Bradford Torrey and
Francis H. Allen, editors. Illustrations. Total of 1888pp. 8⅜ x 12¼.
T312, T313 Two volume set, clothbound $25.00

A SHAKESPEARIAN GRAMMAR, *E. A. Abbott*
Basic reference to Shakespeare and his contemporaries, explaining through
thousands of quotations from Shakespeare, Jonson, Beaumont and Fletcher,
North's *Plutarch* and other sources the grammatical usage differing from the
modern. First published in 1870 and written by a scholar who spent much of
his life isolating principles of Elizabethan language, the book is unlikely ever
to be superseded. Indexes. xxiv + 511pp. 5⅜ x 8½. T1582 Paperbound $2.75

FOLK-LORE OF SHAKESPEARE, *T. F. Thistelton Dyer*
Classic study, drawing from Shakespeare a large body of references to supernatural beliefs, terminology of falconry and hunting, games and sports, good
luck charms, marriage customs, folk medicines, superstitions about plants,
animals, birds, argot of the underworld, sexual slang of London, proverbs,
drinking customs, weather lore, and much else. From full compilation comes
a mirror of the 17th-century popular mind. Index. ix + 526pp. 5⅜ x 8½.
T1614 Paperbound $2.75

THE NEW VARIORUM SHAKESPEARE, *edited by H. H. Furness*
By far the richest editions of the plays ever produced in any country or
language. Each volume contains complete text (usually First Folio) of the
play, all variants in Quarto and other Folio texts, editorial changes by every
major editor to Furness's own time (1900), footnotes to obscure references or
language, extensive quotes from literature of Shakespearian criticism, essays
on plot sources (often reprinting sources in full), and much more.

HAMLET, *edited by H. H. Furness*
Total of xxvi + 905pp. 5⅜ x 8½.
T1004, T1005 Two volume set, paperbound $5.25

TWELFTH NIGHT, *edited by H. H. Furness*
Index. xxii + 434pp. 5⅜ x 8½. T1189 Paperbound $2.75

LA BOHEME BY GIACOMO PUCCINI,
translated and introduced by Ellen H. Bleiler
Complete handbook for the operagoer, with everything needed for full enjoyment except the musical score itself. Complete Italian libretto, with new, modern English line-by-line translation—the only libretto printing all repeats; biography of Puccini; the librettists; background to the opera, Murger's La Boheme, etc.; circumstances of composition and performances; plot summary; and pictorial section of 73 illustrations showing Puccini, famous singers and performances, etc. Large clear type for easy reading. 124pp. 5⅜ x 8½.

T404 Paperbound $1.25

ANTONIO STRADIVARI: HIS LIFE AND WORK (1644-1737),
W. Henry Hill, Arthur F. Hill, and Alfred E. Hill
Still the only book that really delves into life and art of the incomparable Italian craftsman, maker of the finest musical instruments in the world today. The authors, expert violin-makers themselves, discuss Stradivari's ancestry, his construction and finishing techniques, distinguished characteristics of many of his instruments and their locations. Included, too, is story of introduction of his instruments into France, England, first revelation of their supreme merit, and information on his labels, number of instruments made, prices, mystery of ingredients of his varnish, tone of pre-1684 Stradivari violin and changes between 1684 and 1690. An extremely interesting, informative account for all music lovers, from craftsman to concert-goer. Republication of original (1902) edition. New introduction by Sydney Beck, Head of Rare Book and Manuscript Collections, Music Division, New York Public Library. Analytical index by Rembert Wurlitzer. Appendixes. 68 illustrations. 30 full-page plates. 4 in color. xxvi + 315pp. 5⅜ x 8½.

T425 Paperbound $2.25

MUSICAL AUTOGRAPHS FROM MONTEVERDI TO HINDEMITH,
Emanuel Winternitz
For beauty, for intrinsic interest, for perspective on the composer's personality, for subtleties of phrasing, shading, emphasis indicated in the autograph but suppressed in the printed score, the mss. of musical composition are fascinating documents which repay close study in many different ways. This 2-volume work reprints facsimiles of mss. by virtually every major composer, and many minor figures—196 examples in all. A full text points out what can be learned from mss., analyzes each sample. Index. Bibliography. 18 figures. 196 plates. Total of 170pp. of text. 7⅞ x 10¾.

T1312, T1313 Two volume set, paperbound $5.00

J. S. BACH,
Albert Schweitzer
One of the few great full-length studies of Bach's life and work, and the study upon which Schweitzer's renown as a musicologist rests. On first appearance (1911), revolutionized Bach performance. The only writer on Bach to be musicologist, performing musician, and student of history, theology and philosophy, Schweitzer contributes particularly full sections on history of German Protestant church music, theories on motivic pictorial representations in vocal music, and practical suggestions for performance. Translated by Ernest Newman. Indexes. 5 illustrations. 650 musical examples. Total of xix + 928pp. 5⅜ x 8½.

T1631, T1632 Two volume set, paperbound $4.50

THE METHODS OF ETHICS, *Henry Sidgwick*
Propounding no organized system of its own, study subjects every major methodological approach to ethics to rigorous, objective analysis. Study discusses and relates ethical thought of Plato, Aristotle, Bentham, Clarke, Butler, Hobbes, Hume, Mill, Spencer, Kant, and dozens of others. Sidgwick retains conclusions from each system which follow from ethical premises, rejecting the faulty. Considered by many in the field to be among the most important treatises on ethical philosophy. Appendix. Index. xlvii + 528pp. 5⅜ x 8½.
T1608 Paperbound $2.50

TEUTONIC MYTHOLOGY, *Jakob Grimm*
A milestone in Western culture; the work which established on a modern basis the study of history of religions and comparative religions. 4-volume work assembles and interprets everything available on religious and folkloristic beliefs of Germanic people (including Scandinavians, Anglo-Saxons, etc.). Assembling material from such sources as Tacitus, surviving Old Norse and Icelandic texts, archeological remains, folktales, surviving superstitions, comparative traditions, linguistic analysis, etc. Grimm explores pagan deities, heroes, folklore of nature, religious practices, and every other area of pagan German belief. To this day, the unrivaled, definitive, exhaustive study. Translated by J. S. Stallybrass from 4th (1883) German edition. Indexes. Total of lxxvii + 1887pp. 5⅜ x 8½.
T1602, T1603, T1604, T1605 Four volume set, paperbound $11.00

THE I CHING, *translated by James Legge*
Called "The Book of Changes" in English, this is one of the Five Classics edited by Confucius, basic and central to Chinese thought. Explains perhaps the most complex system of divination known, founded on the theory that all things happening at any one time have characteristic features which can be isolated and related. Significant in Oriental studies, in history of religions and philosophy, and also to Jungian psychoanalysis and other areas of modern European thought. Index. Appendixes. 6 plates. xxi + 448pp. 5⅜ x 8½.
T1062 Paperbound $2.75

HISTORY OF ANCIENT PHILOSOPHY, *W. Windelband*
One of the clearest, most accurate comprehensive surveys of Greek and Roman philosophy. Discusses ancient philosophy in general, intellectual life in Greece in the 7th and 6th centuries B.C., Thales, Anaximander, Anaximenes, Heraclitus, the Eleatics, Empedocles, Anaxagoras, Leucippus, the Pythagoreans, the Sophists, Socrates, Democritus (20 pages), Plato (50 pages), Aristotle (70 pages), the Peripatetics, Stoics, Epicureans, Sceptics, Neo-platonists, Christian Apologists, etc. 2nd German edition translated by H. E. Cushman. xv + 393pp. 5⅜ x 8.
T357 Paperbound $2.25

THE PALACE OF PLEASURE, *William Painter*
Elizabethan versions of Italian and French novels from *The Decameron,* Cinthio, Straparola, Queen Margaret of Navarre, and other continental sources — the very work that provided Shakespeare and dozens of his contemporaries with many of their plots and sub-plots and, therefore, justly considered one of the most influential books in all English literature. It is also a book that any reader will still enjoy. Total of cviii + 1,224pp.
T1691, T1692, T1693 Three volume set, paperbound $6.75

THE WONDERFUL WIZARD OF OZ, *L. F. Baum*
All the original W. W. Denslow illustrations in full color—as much a part of "The Wizard" as Tenniel's drawings are of "Alice in Wonderland." "The Wizard" is still America's best-loved fairy tale, in which, as the author expresses it, "The wonderment and joy are retained and the heartaches and nightmares left out." Now today's young readers can enjoy every word and wonderful picture of the original book. New introduction by Martin Gardner. A Baum bibliography. 23 full-page color plates. viii + 268pp. 5⅜ x 8.
T691 Paperbound $1.75

THE MARVELOUS LAND OF OZ, *L. F. Baum*
This is the equally enchanting sequel to the "Wizard," continuing the adventures of the Scarecrow and the Tin Woodman. The hero this time is a little boy named Tip, and all the delightful Oz magic is still present. This is the Oz book with the Animated Saw-Horse, the Woggle-Bug, and Jack Pumpkinhead. All the original John R. Neill illustrations, 10 in full color. 287pp. 5⅜ x 8.
T692 Paperbound $1.75

ALICE'S ADVENTURES UNDER GROUND, *Lewis Carroll*
The original *Alice in Wonderland*, hand-lettered and illustrated by Carroll himself, and originally presented as a Christmas gift to a child-friend. Adults as well as children will enjoy this charming volume, reproduced faithfully in this Dover edition. While the story is essentially the same, there are slight changes, and Carroll's spritely drawings present an intriguing alternative to the famous Tenniel illustrations. One of the most popular books in Dover's catalogue. Introduction by Martin Gardner. 38 illustrations. 128pp. 5⅜ x 8½.
T1482 Paperbound $1.00

THE NURSERY "ALICE," *Lewis Carroll*
While most of us consider *Alice in Wonderland* a story for children of all ages, Carroll himself felt it was beyond younger children. He therefore provided this simplified version, illustrated with the famous Tenniel drawings enlarged and colored in delicate tints, for children aged "from Nought to Five." Dover's edition of this now rare classic is a faithful copy of the 1889 printing, including 20 illustrations by Tenniel, and front and back covers reproduced in full color. Introduction by Martin Gardner. xxiii + 67pp. 6⅛ x 9¼.
T1610 Paperbound $1.75

THE STORY OF KING ARTHUR AND HIS KNIGHTS, *Howard Pyle*
A fast-paced, exciting retelling of the best known Arthurian legends for young readers by one of America's best story tellers and illustrators. The sword Excalibur, wooing of Guinevere, Merlin and his downfall, adventures of Sir Pellias and Gawaine, and others. The pen and ink illustrations are vividly imagined and wonderfully drawn. 41 illustrations. xviii + 313pp. 6⅛ x 9¼.
T1445 Paperbound $1.75

Prices subject to change without notice.

Available at your book dealer or write for free catalogue to Dept. Adsci, Dover Publications, Inc., 180 Varick St., N.Y., N.Y. 10014. Dover publishes more than 150 books each year on science, elementary and advanced mathematics, biology, music, art, literary history, social sciences and other areas.